The SNOWFLAKE

WINTER'S FROZEN ARTISTRY

KENNETH LIBBRECHT
AND
RACHEL WING

Voyageur
Press

First published in 2015 by Voyageur Press, an imprint of Quarto Publishing Group USA Inc., 400 First Avenue North, Suite 400, Minneapolis, MN 55401 USA

Voyageur Press titles are also available at discounts in bulk quantity for industrial or sales-promotional use. For details write to Special Sales Manager at Quarto Publishing Group USA Inc., 400 First Avenue North, Suite 400, Minneapolis, MN 55401 USA.

To find out more about our books, visit us online at www.voyageurpress.com.

ISBN: 978-0-7603-4847-5

Library of Congress Cataloging-in-Publication Data

Libbrecht, Kenneth.
  The snowflake : winter's frozen artistry / by Kenneth Libbrecht and Rachel Wing. -- Revised edition.
    pages cm
  Summary: "A look at what snow crystals are, how they form, different types, their symmetry, and their facets and branches"-- Provided by publisher.
  Includes bibliographical references and index.
  ISBN 978-0-7603-4847-5 (hardback)
  1. Snowflakes--Pictorial works. 2. Snow. I. Wing, Rachel. II. Title.
  QC926.32.L53 2015
  551.57'84--dc23
                          2015011944

Acquiring Editor: Todd R. Berger
Project Manager: Caitlin Fultz
Art Director: Alexandra Burniece
Layout Designer: Simon Larkin

Cover background: Anelina/Shutterstock.com
On the frontis: Diana Kraleva/Moment/Getty Images

Printed in China

10 9 8 7 6 5 4 3 2 1

DEDICATION

To Alanna and Max, our favorite snow-adventure companions,
and to our other family and friends for aiding, abetting, enabling,
cheering—and occasionally affectionately heckling—our snowflake
obsession. Special thanks to Avra for a critical reading of the book.

Contents

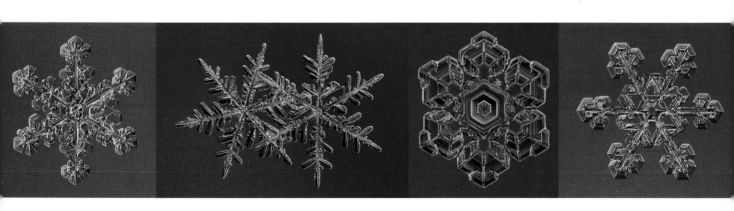

CHAPTER ONE
Winter's Frozen Artistry

"How full of the creative genius is the air in which these are generated! I should hardly admire more if real stars fell and lodged on my coat."
—HENRY DAVID THOREAU, JOURNAL, 1856

"Seriously, you're from Southern California, and you're spending your vacation *here . . . in January . . . on purpose?*" We get this reaction sometimes during our winter travels, as we brave the cold north winds on a quest to find and photograph that most flawless snowflake. And it's true, we have left behind the sunny skies of Pasadena to experience the gray clouds of northern Ontario, Vermont, Michigan, central Alaska, northern Japan, and even the far north of Sweden. Winter has its drawbacks, certainly, but it has charms as well, and few compare with the remarkable beauty one finds in an exquisitely crafted snowflake.

One reason we go trekking through the frozen north is simply that we both enjoy winter, and we want our children to experience it as well. Because we grew up in snow country—Rachel in New York and Ken in North Dakota—our winters were filled with snowballs, snow forts, and snowmen, along with sliding, sledding, and making snow angels. We remember those magical mornings when we would wake up to find a thick blanket of sparkling white covering the landscape, turning our world into a winter wonderland.

Opposite
SNOW CRYSTALS | A sharp eye can distinguish numerous snow stars decorating this branch of an eastern hemlock in Vermont. *Martha Macy*

On snowy afternoons back in elementary school, we especially liked those occasions when the teacher would pass out magnifiers and let us go outside to examine falling snowflakes—the

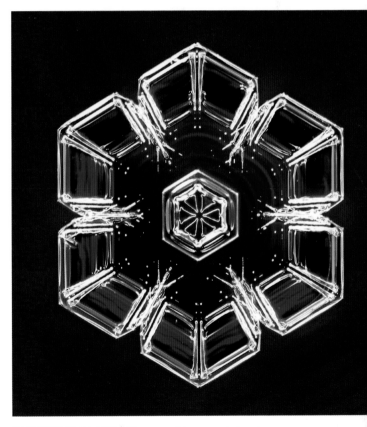

PATTERNS IN ICE | The overall hexagonal structure of this snow crystal is embellished with intricate patterns in the ice.

BRANCHED BEAUTY | Stellar snow crystals like this one have six primary *branches*, and each of these supports numerous *sidebranches*.

ultimate inspiration for our cut-paper creations. The crystals were particularly captivating on colder days, when the starlets would sparkle brightly and linger long enough for a careful inspection of their shape and symmetry. Then the activity became a frenzied treasure hunt as everyone vied to see who could impress the teacher with the largest or most spectacular specimen. If you take time to look closely, there are some amazing constructions falling from the clouds.

Although we both enjoyed our share of snowflakes as children, we nevertheless managed to outgrow their wonder soon after those elementary school days. We lost interest in the icy works of art falling from the sky and no longer paused to ponder how nature crafts such miniature masterpieces. Perhaps the phenomenon was too familiar, to the point that we simply stopped noticing. Or maybe we just had too much of a good thing; it can be difficult to appreciate the refined beauty of snowflakes when the driveway is piled high with them and you have a shovel in your hand.

It was only much later in life—many years after relocating to Southern California—that we began to develop a deeper appreciation for the diminutive snowflake. Ken had begun researching the science of how crystals grow, and his attention turned one day to the physics

In the pages that follow, we journey into the winter clouds, watch snowflakes as they are born and grow, and examine the origin of their form and symmetry. We grow our own snowflakes in the laboratory, where we can watch the process as it happens, and create a few exotic varieties under conditions not found in nature. We consider how these remarkable, six-fold symmetrical structures assemble themselves, quite literally out of thin air. We attempt to answer that seemingly simplest of children's questions: where do snowflakes come from?

So put on your snow boots, pick up your magnifying glass, and let us be your guides as we take an up-close look at winter's frozen artistry.

ENDLESS VARIATIONS | Snow crystals fall to earth in an endless variety of remarkable shapes and patterns.

of ice and the formation of those long-forgotten snowflakes. Perhaps his North Dakota roots were exerting their influence, but the science focus put new eyes on the subject. Rachel was drawn in as well, and soon we both began seeing more than we ever saw growing up.

A simple magnifier reveals a remarkable diversity of different snowflake types, and viewing individual crystals under a microscope opens up a world of amazingly intricate structures. How extraordinary it is that such beautifully complex forms can simply appear, spontaneously, gently falling to earth in vast, vast numbers.

Snowflake science soon led to snowflake photography, and before long we were on a train heading north toward Hudson Bay in the middle of January, children in tow. Ten thousand pictures later, we hope to share with you some striking snowflake images and the story of how snowflakes are created.

FACETS IN ICE | Many snow crystals exhibit mirror-like faceted surfaces that call to mind the look of a cut gemstone.

WHAT IS A SNOWFLAKE?

When we say *snowflake*, we usually mean *snow crystal*. The terms are often used synonymously, but there is a distinction in their meteorological definitions. A *snow crystal*, as the name implies, refers to a single crystal of ice, within which the water molecules are all lined up in a precise hexagonal array. Whenever you see that characteristic six-fold symmetry, you know you are looking at a snow crystal.

A *snowflake*, on the other hand, is a more general term that can mean an individual snow crystal, a cluster of snow crystals that form

FERNLIKE FORMS | Closely spaced sidebranches give these snowflakes a leafy appearance reminiscent of a fern.

together, or even a large aggregate of snow crystals that collide and stick together in mid-flight. Those large puffballs you see floating down in warmer snowfalls are called snowflakes, and each is made of hundreds or even thousands of individual snow crystals. Snow crystals are commonly called snowflakes, and this is fine, like calling a tulip a flower.

Many people think snowflakes are simply frozen raindrops, but this is not true. Raindrops do sometimes freeze in midair as they fall, but this type of precipitation is called *sleet*. Sleet particles look like what you might expect—drops of frozen water without any of the delicate patterning or symmetry seen in snowflakes. A snowflake appears when water vapor in the air converts directly into ice without first becoming liquid water. As more vapor condenses onto a nascent snow crystal, it grows and develops, and that is when its ornate patterning emerges.

EPHEMERAL ART | Snowflakes are best photographed immediately after their descent from the clouds. Once inside a snowbank, intricately patterned snowflakes slowly lose their finer features.

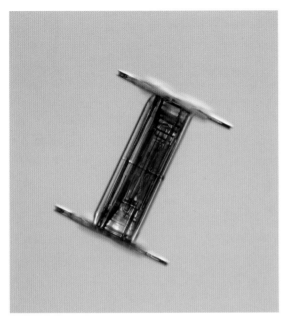

CAPPED COLUMNS | These two photographs show *capped columns*—a little-known type of snowflake that appears when two stellar-plate crystals grow out from the ends of an ice column.

The canonical snowflake is an elaborate, six-branched snow star. These are the ever-popular icons of ski sweaters and winter-holiday decorations. Nature produces a great many variations of this type of snow crystal, each exhibiting its own unique shape. The six primary branches may support secondary *sidebranches*, sometimes so numerous that the crystals have a leafy, almost fernlike appearance. Sometimes the sidebranching is quite symmetrical, but often it is not.

One thing you will not find in nature is a bona fide eight-sided snow crystal. The same is true of four-, five-, and seven-sided snow crystals. The symmetry of the underlying molecular lattice does not allow such forms. Eight-sided snowflakes may be easier to cut out of paper, but real snow crystals never have eight-fold symmetry, regardless of what you see in holiday decorations.

EXOTIC SNOWFLAKES

A snowflake is a temporary work of art. To capture most of the images in this book, each snowflake was plucked from the air as it fell and then rapidly photographed. In mere minutes, a fallen snowflake starts to lose its shape. The sharp corners begin to round, and after a brief time many of its delicate features are gone. No two snowflakes look exactly alike when they fall, but their uniqueness is soon lost when they sit on the ground. Inside a snowbank, intricately patterned snowflakes slowly transform into tiny lumps of ice. To see the best snowflakes, you have to catch them when they are fresh.

When we first began reading about snowflakes, our knowledge extended only as far as six-branched snow stars. These are the largest snow crystals, making them the easiest to see on your sleeve; plus, they are pretty much the only snowflakes found in popular media. We were quite surprised when our reading uncovered a whole menagerie of different snowflake types, of which the stellar snow crystal is just one example.

We were especially struck when we came across old photos of what is called a *capped column*—a short ice column with plate-like crystals on each end, resembling two wheels on an axle or an empty spool. Neither of us had ever witnessed such a peculiar construction growing up. As we began looking for them, however, capped columns were not difficult to find. We had overlooked them in our youth simply because we were oblivious to their existence.

If you live in snow country, there is a good chance that you too are living unaware of the exotic capped column. If so, please have a look for yourself next time it snows. You may not see this noteworthy ice creation in every snowfall, but capped columns, as well as many other snowflake varieties, are out there if you know what to look for.

Columns and needles are also common snow crystal forms. Small in size, they are often present but seldom noticed. Their most basic shape is a simple hexagonal column of ice, similar to the shape of a standard wooden pencil. Often the columns have hollow ends, while at other times they may grow into assemblies of ice needles. A snowstorm will occasionally bring great numbers of these small rods of ice, resembling short bits of white hair when they land on your sleeve.

The snowflake menagerie includes a great many variations on the hexagonal theme, from slender columns to thin plates, at various times branched, sectored, patterned, hollowed, and faceted. Different weather conditions yield different kinds of snowflakes. Large stellar crystals appear on colder days, while columnar forms are common just below freezing. Often the crystals change during the course of a snowfall, as the clouds evolve. There are all sorts of complex and curious designs out there waiting to be seen, all floating lazily to earth. One does not easily become bored looking at snowflakes.

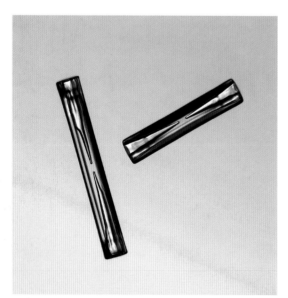
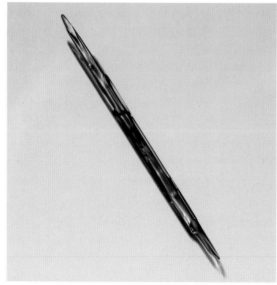

COLUMNS AND NEEDLES | Columnar crystals (left) and ice needles (right) are common snowflake forms.

THE SECRET LIFE OF A SNOWFLAKE

Although the snowflake design possibilities are endless, that does not mean just any shape or pattern can be seen falling from the sky. The patterns that emerge are not arbitrary, but rather they are determined by the processes governing snow crystal formation. When you know how to interpret their icy architecture, the shape of a snow crystal tells a story about how it was created.

Snowflakes are not made by machines, nor are they alive. There is no blueprint or genetic code that guides their construction. Yet they appear in these amazingly ornate, symmetrical shapes. Snowflakes are simple bits of frozen water vapor, flecks of ice that tumble down from the clouds. How do they develop into such intricate six-petaled stars? What subtle forces are responsible for engineering the never-ending variety of snow crystal structures?

In a snowflake, just an ordinary snowflake, we can find a fascinating tale of the spontaneous emergence of pattern and form. From shapeless water vapor, complex structures emerge in an airborne symphony of meteorological morphogenesis. Snowflakes are the product of a rich synthesis of physics, mathematics, and chemistry—and they're fun to catch on your tongue.

TWELVE-BRANCHED SNOWFLAKE | Two six-branched crystals sometimes stick together to form a twelve-branched snowflake.

The photograph above shows small piles of salt, sugar, and crushed glass (from left to right). All three of these materials are intrinsically colorless and transparent (glass especially so), but the piles look white because they reflect light instead of absorbing it. Incident light reflects off all the tiny surfaces of the grains, and the sum of these many reflections makes the piles look white. A snowbank looks white for the same reason, as does a cloud. Ice and water are essentially clear and colorless, but the surface reflections from countless particles produce a white appearance.

An individual snow crystal seen close up does not look white but clear, like a small sliver of shaped glass. When photographing snowflakes, we typically illuminate each crystal from behind, so the light is transmitted through the ice. The clear material bends the rays of light, giving the photograph a sense of depth and accentuating each crystal's internal structure and patterning. We often illuminate using colored lights as well, as this highlights structural features and can create a variety of pleasing photographic effects.

nycshooter/Vetta/Getty Images

CHAPTER TWO
Snowflake Watching

"No weary journeys need be taken, no expensive machinery employed. . . . A winter's storm, an open window, a bit of fur or velvet, and a common magnifier, will bring any curious inquirer upon his field of observation with all the necessary apparatus, and he has only to open his eyes to find the grand and beautiful laboratory of nature open to his inspection."
—FRANCES CHICKERING, *CLOUD CRYSTALS: A SNOW-FLAKE ALBUM*, 1864

The snowflake made its first appearance in recorded history when people identified individual snow crystals—with their distinctive six-fold symmetry—as the constituent elements of falling snow. The earliest known account was in 135 BC, when Chinese scholar Han Yin wrote, "Flowers of plants and trees are generally five-pointed, but those of snow, which are called ying, are always six-pointed." Subsequent Chinese writers mentioned snowflake symmetry as well, an example being the sixth-century poet Hsiao Tung, who penned: "The ruddy clouds float in the four quarters of the cerulean sky. And the white snowflakes show forth their six-petaled flowers."

European authors began documenting snowflakes many centuries later; the first known reference was from Scandinavian bishop Olaus Magnus in 1555. However, the clergyman described snowflakes as having a peculiar assortment of shapes, including crescents, arrows, bells, and even a human hand, so perhaps his account does not quite warrant being called a historical first. English astronomer Thomas Harriot did a better job, correctly identifying the snowflake's six-fold symmetry in 1591.

Although Europe arrived late into the snowflake story, the Renaissance was revitalizing intellectual life by the end of the sixteenth century. European science ramped up quickly as scholars began scrutinizing any and all natural phenomena with renewed vigor and an eye for mathematical precision. Their scientific curiosity was soon piqued as they took a closer look at falling snowflakes.

Opposite
SNOW STAR | We found this magnificent snow crystal in Burlington, Vermont.

SNOWFLAKE SCRUTINY

French philosopher and mathematician René Descartes recorded the first accurately detailed account of snow crystal structure in 1637. In his scientific study of meteorology and weather, *Les Météores*, Descartes recorded some remarkably thorough naked-eye observations of snow crystals, which included several of the rarer forms:

> After this storm cloud, there came another, which produced only little roses or wheels with six rounded semicircular teeth . . . which were quite transparent and quite flat . . . and formed as perfectly and symmetrically as one could possibly imagine. There followed, after this, a further quantity of such wheels joined two by two by an axle, or rather, since at the beginning these axles were quite thick, one could as well have described them as little crystal columns, decorated at each end with a six-petaled rose a little larger than their base. But after that there fell more delicate ones, and often the roses or stars at their ends were unequal. But then there fell shorter and progressively shorter ones until finally these stars completely joined, and fell as double stars with twelve points or rays, rather long and perfectly symmetrical, in some all equal, in others alternately unequal.

In this passage, we can see snowflakes influencing—in their own small way—the early development of what is now modern science. Descartes was clearly impressed with the geometrical perfection he saw in snow crystal forms, with their flat facets and hexagonal symmetry. Pondering this and other observations, he went on to reason how the principles of geometry and mathematics play a central role in describing the natural world.

EARLY SNOWFLAKE SKETCHES | These snowflake drawings were published in 1665 by English scientist Robert Hooke in his famous book focusing on the miniature world, *Micrographia*. These were the first observations of snowflakes derived from a magnified view.

METICULOUS SNOWFLAKE OBSERVER | French philosopher and mathematician René Descartes is perhaps best known for his metaphysical dictum *Cogito, ergo sum* (I think, therefore I am). He also made some of the most thorough early observations of snow crystals. *Frans Hals—Portret van René Descartes*

SNOWFLAKE INSPIRED KIMONO | This nineteenth-century woodblock by Japanese artist Utagawa Kunisada shows a fashionable woman sporting snowflake motifs on her kimono. It was most likely inspired by drawings of snowflakes published in 1832 by Toshitsura Doi. *Image copyright © The Metropolitan Museum of Art. Image source: Art Resource, NY*

Although we take this for granted now, using mathematics to explain ordinary phenomena was still an emerging idea at the time and a major step forward in science.

The invention of the microscope in the mid-seventeenth century quickly led to more and better snowflake observations. English scientist and early microscopist Robert Hooke sketched snowflakes and practically everything else he could find for his book *Micrographia*, published in 1665, which quickly became the world's first scientific bestseller. Although his microscope was crude by modern standards, Hooke's drawings nevertheless began to reveal the complexity and intricate symmetry of snow crystal structure, details that could not be detected with the unaided eye. As the quality and availability of optical magnifiers improved, so did the accuracy of snow crystal drawings. By the mid-nineteenth century, a number of observers around the world had documented the diverse character of snow crystal forms.

ARCTIC SNOW CRYSTALS | English explorer William Scoresby made these sketches during a winter voyage through the Arctic, which he recounted in his 1820 book, *An Account of the Arctic Regions with a History and Description of the Northern Whale Fishery*. They are the first drawings that accurately diagrammed many features of snow crystal structures, as well as several rare forms, including triangular crystals and capped columns. Scoresby also noted that the cold arctic climate produces more highly symmetrical crystals than typically seen in Britain.

FIRST SNOWFLAKE PHOTOGRAPHY

Wilson Bentley, a farmer from the small town of Jericho, Vermont, created the first photographic album of falling snow, which awakened the world to the hidden wonders of snowflakes. As a teenager in the 1880s, Bentley became interested in the microscopic structure of snow crystals, and he began experimenting with the new medium of photography as a means of recording what he observed. He constructed an ingenious mechanism for attaching a camera to his microscope for this purpose, and he succeeded in photographing his first snow crystal in 1885, when he was nineteen years old.

To say Bentley was dedicated to snowflake photography is an understatement. The winter pastime became his lifelong passion, and over the course of forty-six years he captured more than five thousand snow crystal images on glass photographic plates. He resided in the same Jericho farmhouse for his entire life, photographing snowflakes each winter using the same equipment he constructed as a teenager.

Bentley's photographs appeared in numerous publications over several decades, providing for many their first look at the inner structure and symmetry of snow crystals. And with thousands of snowflakes, all unique, the world was exposed

FIRST PAPER SNOWFLAKES | Frances Chickering, a minister's wife from Maine, published these snow-crystal images in her 1864 book *Cloud Crystals: A Snow-Flake Album*. She examined snow crystals as they fell on her windowsill and quickly cut paper replicas of their forms, which were later transferred to the pages of her album. Ms. Chickering could hardly have imagined that the craft she was pioneering would later be practiced by millions of school children around the world.

to their incredible variety as well. The now-familiar old chestnut that no two snowflakes are exactly alike appears to have had its origin in Bentley's photographs.

In the late 1920s, Bentley teamed with W. J. Humphreys, chief physicist for the United States Weather Bureau, to author his magnum opus, containing more than two thousand snow crystal photographs. Alas, the sixty-six-year-old Vermont farmer died of pneumonia just a few weeks after the work was published.

In the decades that have followed this seminal work, many others have taken up the challenge of capturing the structure and beauty of snow crystals using ever-improving cameras and lenses. We like to think that Wilson Bentley would smile and give a knowing nod of approval if he could see how snowflake photography has flourished over the years.

CAPTURING SNOWFLAKES | Vermont farmer Wilson Bentley developed the art of snowflake photography in the late 1800s, eventually producing a large album of images. He is shown here with his specially built snow-crystal photo-microscope. *Jericho Historical Society*

FIRST SNOWFLAKE PHOTOGRAPHS | These are just a few of the thousands of snowflake photographs taken by Wilson Bentley between 1885 and 1931. The original photos showed a bright crystal against a bright background, since he illuminated his snowflakes from behind. The photos were modified by essentially cutting each crystal out and placing it on a dark background. *Jericho Historical Society*

LABORATORY SNOWFLAKES

The snowflake story took a decidedly scientific turn in the 1930s, as Japanese physicist Ukichiro Nakaya began the first laboratory investigation of snow crystals. While earlier snowflake watchers had been limited to fairly basic observations, Nakaya brought to bear the arsenal of new scientific tools that had become available at the beginning of the twentieth century.

During his education as a promising young experimental physicist, Nakaya studied electrical discharges and x-ray generation, both cutting-edge topics of the day. Finding a job was difficult in the early 1930s, however, and

fate took Nakaya to the northern reaches of Japan, where he landed a position at Hokkaido University. Research funds and equipment were minimal at the university, but the cold climate provided an ample supply of snow crystals, and Nakaya quickly became interested in their study.

Inspired by Bentley's photographs, Nakaya began his own photographic investigations, classifying different snow crystal types along with the atmospheric conditions that led to their formation. He soon discovered that Bentley's dazzling stellar crystals were just the beginning. Nakaya added many detailed observations of columns, needles, capped columns, and other less common forms, thus producing the first photographic documentation of the broader menagerie of falling snow.

Having learned a great deal from direct observations, Nakaya then turned his attention to better understanding how these icy structures were created in the clouds. As he had become a world expert on winter precipitation, the scientist secured funding to build a walk-in freezer laboratory at Hokkaido, exploiting the recently invented technology of refrigeration. With this new facility, Nakaya set out to create his own snowflakes for closer scrutiny and study.

Growing snow crystals in the laboratory turned out to be a considerable challenge. Natural snowflakes float freely in the atmosphere as they develop, and their long descent gives them ample time to grow to a substantial size. Because a mile-high freezer was clearly impractical, Nakaya needed a different approach to produce crystals that were large enough to study.

Nakaya sought to suspend an individual snow crystal on a fine string, where he could watch it grow and develop. That way the growth time could be extended indefinitely. But here he hit a snag. He wanted an isolated snow crystal, but what he got was a large number of tiny ice crystals that coated his string with frost. Nakaya explored many different filaments in his quest to make solitary snow crystals, including silk, cotton, fine wires, and even a spider's web. All resulted in frost-like clusters of minute ice crystals.

Nakaya finally achieved success in 1936 with, of all things, a rabbit hair. The natural oils on the hair discouraged ice nucleation and prevented the growth of large numbers of frost crystals. Instead, isolated snow crystals grew into forms that bore an excellent resemblance to those produced in the clouds. These were the world's first synthetic snowflakes.

We use the word *synthetic* here because Nakaya produced real snow crystals, grown from water vapor in the same manner as the natural variety. They were synthesized in the laboratory but otherwise were the same as what appears in a real snowfall. This is in contrast to the *artificial* snow now made at ski resorts, which is not made from water vapor. Artificial snow is made by quickly freezing liquid water droplets (as we will describe in a later chapter), yielding small sleet particles that are markedly different from natural snowflakes.

THE MORPHOLOGY DIAGRAM

After discovering the unexpected virtues of rabbit hair, Nakaya spent years growing many individual snow crystals at different temperatures and humidity levels. He observed how the morphology of each crystal—its detailed shape and structure—depended on the condition of the air in which it grew. He combined all these observations into what is now called the *snow crystal morphology diagram*, or often the *Nakaya diagram*.

The morphology diagram is like a Rosetta stone for snowflakes. With it, you can translate the shape of a falling snowflake into a description of its growth history. Upon seeing a slender needle crystal, for example, you know that it grew in high humidity at a temperature near 21 degrees Fahrenheit (−6 °C). A large stellar crystal suggests growth near 5 degrees Fahrenheit (−15 °C), while extravagant sidebranching indicates high humidity.

The morphology diagram is especially illuminating when growth conditions change as a snow crystal forms, a good example being the capped

THE SNOW CRYSTAL MORPHOLOGY DIAGRAM

With the snow crystal morphology diagram, one can infer the temperature and humidity in the winter clouds simply by looking at the falling snowflakes. Temperature mainly determines whether snow crystals grow into platelike or columnar forms. Platelike crystals appear when the temperature is near either 28 degrees Fahrenheit (−2 °C) or 5 degrees Fahrenheit (−15 °C), as seen in the diagram. Especially large, well-formed stellar snow crystals usually appear only when the temperature is around 5 degrees Fahrenheit (−15 °C). Columnar and needlelike crystals are most likely to occur when the temperature is around 21 degrees Fahrenheit (−6 °C).

The humidity level influences the complexity of the growing crystals. When the humidity is low, snow crystals grow slowly into simpler, faceted shapes. At higher humidity levels, crystals grow rapidly into complex, branched structures.

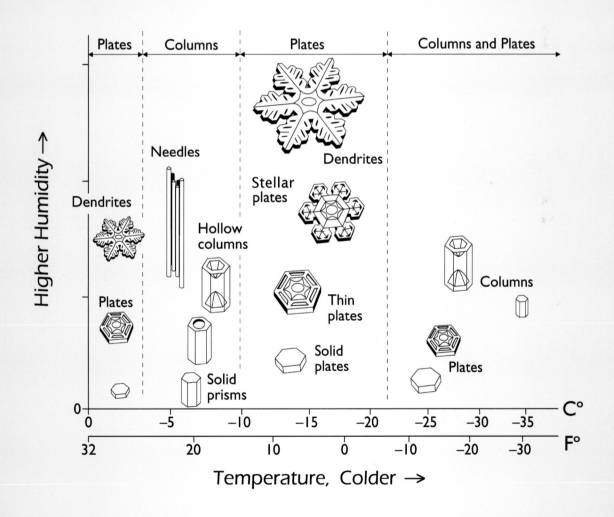

HIEROGLYPHS FROM THE SKY | The shape of a snowflake tells a story about the changing temperature and humidity it experienced while it was growing.

column. This structure appears when a crystal initially begins growing in a region of the clouds where the temperature is around 21 degrees Fahrenheit (–6 °C), developing into a columnar crystal. Then the wind takes it to a region where the temperature is closer to 5 degrees Fahrenheit (–15 °C), and at that point plates begin growing on the ends of the column. The morphology diagram provides a straightforward explanation for the origin of capped columns and other complex crystal types.

Nakaya used to remark that snowflakes are like hieroglyphs from the sky. With the morphology diagram, one can connect the shape of a snow crystal to the atmospheric conditions that generated it. The structure of an individual snow crystal can be deciphered, as if a kind of meteorological hieroglyphics, to reveal the conditions of the clouds in which it formed.

The morphology diagram does even more: it helps explain the extreme diversity of snow crystal shapes. In particular, the diagram reveals that even a few degrees of temperature change can dramatically alter a crystal's growth behavior. As it moves through the clouds, small changes in temperature and humidity affect its development in big ways, yielding complex crystal shapes. We will come back to the morphology diagram many times in this book, as the diagram has become an essential tool for understanding the variable nature of snow crystal formation.

FAMILY FUN

We rediscovered the pleasures of snowflake watching in late 1996 during a Christmas vacation visiting family in the town of Grafton, in the northeast corner of North Dakota. We had not yet become true snowflake enthusiasts, but we had brought along a copy of Nakaya's classic 1954 text, *Snow Crystals: Natural and Artificial*. We had recently acquired the book from a rare-book dealer, and it looked to be an interesting (albeit unusual) vacation read. Thus began our utter fascination with the subject, and that winter holiday became something of a turning point in our study of snowflakes.

Nakaya describes a variety of different types of snow crystals in his book, and one of these is the capped column, which he calls a *Tsuzumi* crystal, after a double-sided Japanese drum having a similar shape. Because neither of us had ever encountered this peculiar snow crystal before, our initial thought was that capped columns must be terribly rare and difficult to find. After all, we had seen plenty of stellar snowflakes growing up but nothing resembling Nakaya's *Tsuzumi* crystal.

A few days later we were out for a walk with the kids on Park River, which makes its way through Grafton as it meanders toward the Red River of the North. And by "walking on the river," we do not mean *along* the river; this was North Dakota in the middle of winter, so the rivers were all frozen over and made fine walking trails. Our kids were little back then, and they had never experienced so much snow before. They were having a delightful time playing on the river, in spite of the cold.

A light snow began falling as we were on this outing, and Rachel, showing excellent foresight, had brought with her a small, fold-up magnifier. We quickly discovered a peculiar law of nature that seems to place the most interesting snowflakes on other people's clothing. It must have been an odd sight—the four of us milling about in the middle of a frozen river, scrutinizing each other's coats and stocking caps with great interest.

At first we saw a number of small hexagonal crystals and some stellar plates, and with continued looking we found quite a few branched stars. From time to time we would find stellar crystals that were quite large and ornate. But then, wait, what was that peculiar shape? It looked like a capital "I" at first glance, seeing it from the side, and then we suddenly realized what we had found. It was our first capped column.

ENJOYING SNOWFLAKES

If you happen to live in snow country, or if you travel to snowy places from time to time, then we heartily recommend that you take a close look at the falling snow whenever you have a chance. Nature quietly provides billions upon billions of exquisite ice sculptures that simply float down from the clouds, all impermanent, nearly all unnoticed. Each snowfall provides a unique exhibit of winter art, and the show deserves an audience.

As a winter recreation, snowflake watching is delightfully simple and fun for everyone. It requires the most minimal investment in time or equipment. You need little more than an inexpensive magnifier and a gentle snowfall. If the clouds do their part, then you too might be struck by the sight of a magnificent snow star, a twelve-branched snowflake, or your first capped column. As Yogi Berra once commented, "You can observe a lot just by watching."

Not every snowfall brings magnificent snowflakes. At times, especially when it's relatively warm, a storm will bring little more than gloppy clumps that offer little for study. The crystals melt a bit in flight, or they collide and stick together before hitting the ground. As Henry David Thoreau observed, "commonly the flakes reach us travel-worn and agglomerated, comparatively without order or beauty, far down

in their fall, like men in their advanced age." When all we see fits this description, we may just put away our magnifiers and try again some other time.

The finest snowflake watching seems to be reserved for those cold, quiet, gray winter days. The need for cold follows directly from the morphology diagram, which tells us that 5 degrees Fahrenheit (−15 °C) is the ideal temperature for finding large and ornate stellar crystals. Beyond that, however, it is difficult to predict what will fall from the clouds, and a light flurry of spectacular crystals can easily go unnoticed. No trumpeters will announce the most exquisite flakes of frozen art. You simply have to be vigilant so that you notice when they arrive.

If you have any plans for sledding, skiing, snowmobiling, or perhaps walking down the banks of a frozen river, remember to bring along your magnifier. If children will be present, then we suggest quietly packing a magnifier and bringing it out when the time is right, when you notice some eye-catching crystals on your sleeve. Children are born curious, and they love looking at snowflakes.

On a day when the crystals are falling well, you will be amply rewarded for your diligence. You may even attract a crowd of interested onlookers, on the ski slope or wherever, wondering what the fuss is about. Although a simple magnifying lens cannot reveal all the minute details in a snowflake, there are many remarkable sights you can find on your coat sleeve. Photographs cannot convey the sparkle you see as you move a crystal around to observe the light playing on its facets. Add to that the quiet pleasure of watching the flakes float lazily to earth, plus the excitement of discovering exceptionally well-formed crystals, and you have a winter recreation that should not be missed.

TREASURE HUNTING | You never know what you might find when hunting for snowflakes. Some display ornate patterns and elaborate branching, while others have an elegant simplicity with mirror-like facets. Each snowfall brings something different.

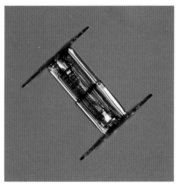

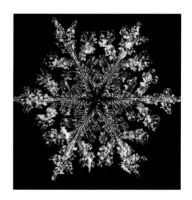

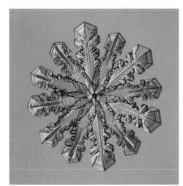

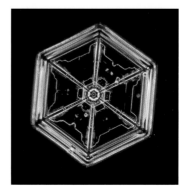

SNOWFLAKE FOSSILS

Preserve a snow crystal in resin and you can share it with friends and family in the comfort of your living room—your very own personal snowflake fossil. We made these two examples using superglue, and you can easily do this yourself. You will need a few inexpensive supplies:

1) A tube or two of thin liquid superglue—not the gel type—which you can find at your local drugstore or hardware store.
2) Some glass microscope slides and cover slips, which are readily available online, often both together in a box for less than $10.
3) A small paintbrush, the smaller the better.
4) A piece of dark cardboard or dark cloth.

Pre-chill all these items outside, for example in an unheated garage, a backyard shed, or just under an eave. When snow is falling, let some crystals land on the cardboard and have a look at what nature is providing. Use your magnifier for a better view. When you see an outstanding crystal, gently pick it up with the paintbrush (it's easier than it sounds) and place it on a clean glass slide. Carefully add a good-sized drop of cold superglue right on top of the crystal, and place a cover slip on top of that. As you can see in the examples on these two pages, you may trap a few air bubbles along with your snowflakes in the superglue.

All this has to be done outside in the cold, and your supplies all have to be cold as well. You have to be careful that heat from your hands doesn't melt the crystal, and even just breathing on your snowflake can be detrimental. If you find that the crystals are melting no matter what you do, then you may just have to wait for a colder day. Handling snowflakes is generally easier in colder conditions; it can be downright frustrating when the temperature is just below freezing.

Finally, let the glue harden while it remains at sub-freezing temperatures, which may take up to a week. If the weather is going to stay cold, you can leave your slides outside in a sheltered spot. If not, pop them in your kitchen freezer. After the glue hardens, the slides can be brought to room temperature. Your fossilized snowflakes, like insects trapped in amber, should last indefinitely.

CHAPTER THREE
A Field Guide to Falling Snow

"Some of Nature's most exquisite handiwork is on a miniature scale, as anyone knows who has applied a magnifying glass to a snowflake."
—RACHEL CARSON, *SENSE OF WONDER*, 1956

Snow crystals can be found in a marvelous diversity of sizes and shapes. Some forms are common, appearing in most snowfalls, while a number of exotic types are seen only infrequently. What you find depends on the weather conditions in the clouds, especially the temperature and humidity surrounding each crystal as it grows. In this chapter we take a brief tour of the snowflake menagerie, examining the different crystal types and under what conditions you might find them.

DIAMOND DUST

We start our tour with the smallest snow crystals, known in the aggregate as *diamond dust*. Diamond-dust snow crystals are typically only a few tenths of a millimeter in size, so tiny that a dozen or more might fit inside a single poppy seed. Individual crystals are nearly invisible to the naked eye, but you can see them sparkle in sunlight or under a bright street lamp. On especially chilly days with no wind, the crystals drift slowly through the air around you like glittering dust particles, and this is where they got their name.

These are the simplest snow crystals, their basic form being a *hexagonal prism*, defined by two *basal facets* and six *prism facets*. Variations

run the gamut from long, slender hexagonal columns to thin, broad hexagonal plates. Most snow crystals begin their existence as diamond dust, developing their more elaborate branched patterns as they grow larger.

Diamond-dust crystals are common up in the clouds but less so near the ground. The diminutive specks of ice are often so small that they can remain suspended in the air indefinitely, just like cloud droplets. Larger diamond-dust crystals are most common in Arctic regions, as frigid temperatures often yield simpler prisms with flat facets that sparkle brightly. In warmer regions, the crystals are usually more rounded or patterned, so their sparkle is not as vibrant.

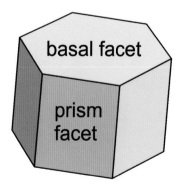

Opposite
MAGNIFICENT ICE | This stellar snow crystal grew slowly, forming intricate surface markings.

HEXAGONAL PRISM | This most basic snow-crystal shape is defined by eight planar surfaces—two basal facets and six prism facets.

THE SIMPLEST SNOWFLAKES | In temperate climates, diamond-dust crystals are basically hexagonal prisms, but most have some additional patterning.

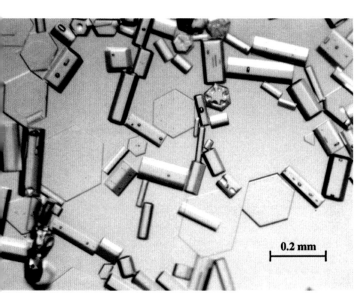

ANTARCTIC SNOWFLAKES | Sharply faceted prisms are most often seen in extremely cold climates. These diamond-dust crystals were collected at the South Pole. *Walter Tape*, Atmospheric Halos

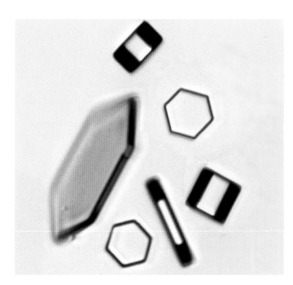

SIMPLE CRYSTALS | These tiny, laboratory-grown snow crystals, no larger than the diameter of a human hair, are simple hexagonal prisms.

LAVISH BRANCHING | Fernlike stellar dendrites are a common snow crystal form, and their large size makes these snowflakes easy to recognize.

STELLAR DENDRITES

Jumping from the smallest to the largest, next are branched snow stars, or *stellar dendrites*. The word dendrite derives from the Greek word for tree, and these snow crystals display six main branches flanked by additional sidebranches. When the sidebranching is especially extravagant, the crystals resemble lacy garden ferns, so these are called *fernlike stellar dendrites*. If a stellar crystal has no sidebranches at all, we call it a *simple star*.

Stellar dendrites fall with sizes up to around one-quarter of an inch (about 6 millimeters)—occasionally even larger. You can spot them easily on your coat sleeve, and you get a clear view using just a simple magnifier. Although large in diameter, stellar dendrites are surprisingly thin and flat; a large specimen might be one hundred times wider than it is thick. These crystals put the "flake" in snowflake.

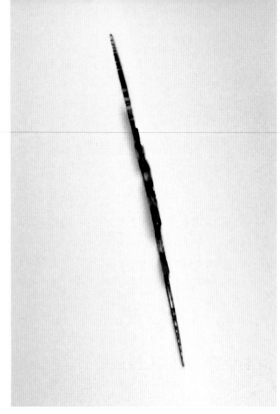

Stellar dendrites are among the most common of the snow crystal types. They form when the temperature is near 28 degrees Fahrenheit (−2 °C) and also when it is near 5 degrees Fahrenheit (−15 °C), as seen in the snow crystal morphology diagram. At times, a snowfall might drop almost entirely stellar dendrites. If there is little wind, the thin flakes drift lazily down from the clouds in a most unhurried fashion, embracing the landscape in a serene snowfall.

Snowflake photographers are always happy to see stellar dendrites falling from the clouds, as nature brings her full artistic talents to bear when making these showy crystals. Their intricate branched patterns make stellar dendrites a feast for the eye and an excellent subject for the camera.

Fernlike stellar dendrites are the fastest-growing snow crystals, reaching full size in as

PAPER THIN | Viewing stellar dendrites from the side reveals that they are remarkably thin and platelike.

ORNATE PATTERNS | Stellar dendrites can be found in a marvelous variety of shapes and sizes.

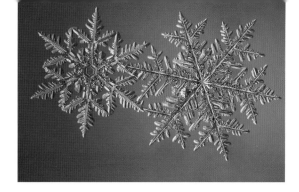

little as fifteen minutes. Their rapid growth drives the formation of many irregularly spaced sidebranches. Although it has a complex structure, a fernlike stellar dendrite is still a single crystal of ice, so the water molecules are all lined up from tip to tip. As a result, each sidebranch grows outward at a 60-degree angle from its main branch, and the sidebranches are parallel to neighboring primary branches.

Powder skiers are especially happy to see a storm dropping lots of fernlike stellar dendrites. The barbed branches and light weight of these crystals mean they pack loosely together when they land. If the resulting snowpack is light enough, it may warrant being called "champagne powder." This is the kind of effervescent snow that gives skiers the feeling that they are floating down the slopes.

CHAMPAGNE POWDER | Snowfalls often produce almost entirely stellar-dendrite crystals, which pack loosely together when they land. Fernlike stellar dendrites provide the best powder skiing.

ORDER AND RANDOMNESS | Each sidebranch in these fernlike stellar dendrites is parallel to its neighboring main branch, while the spacing between sidebranches is irregular.

DISTINCTIVE PATTERNS | Endless variation is the hallmark of stellar dendrites.

Opposite

MONSTER SNOW CRYSTAL | To our knowledge, this is the largest snow crystal ever photographed—a fernlike stellar dendrite measuring just over 10 mm (0.4 inches) from tip to tip. We have witnessed such large crystals only twice, both times in Cochrane, Ontario, and both times for just a few short minutes. Each branch holds first-generation sidebranches along with second-, third-, and even fourth-generation sidebranches. Higher-order sidebranching like this is rare in snow crystals.

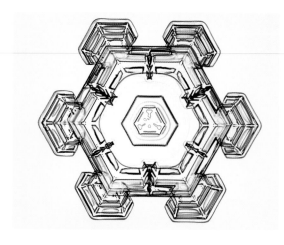

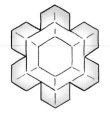

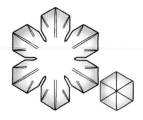

STELLAR PLATES

When stellar crystals exhibit little or no sidebranching, they are called *broad-branched stellar plates* or just *stellar plates*. These crystals are typically smaller than stellar dendrites, but they grow at the same temperatures, around 28 degrees Fahrenheit (–2 °C) or 5 degrees Fahrenheit (–15 °C). As can be seen in the morphology diagram, the humidity level determines the amount of sidebranching in plate-like crystals. Simple plates and stellar plates appear at the lowest humidity levels, progressing to stellar dendrites and finally fernlike stellar dendrites as the humidity increases.

While stellar plates have a simpler outline than stellar dendrites, they often exhibit elegant surface patterning. Sometimes the ridges, grooves, and other surface features exhibit an extraordinary degree of six-fold symmetry that is not found in other snow crystal types.

Stellar plates sometimes include broad, smooth regions that are seemingly divided into sectors by ridges in the ice, and these crystals are called *sectored plates*. The simplest sectored plate is an otherwise featureless hexagonal plate divided into six parts, like equal slices of a hexagonal pie.

The dividing line between different snow crystal types is rarely sharp, and the sectored

INTRICATE PATTERNS | Stellar plates are often embellished with striking surface markings.

Opposite
SIZE COMPARISON | This composite image shows the sizes of some snow crystals in comparison to a penny.

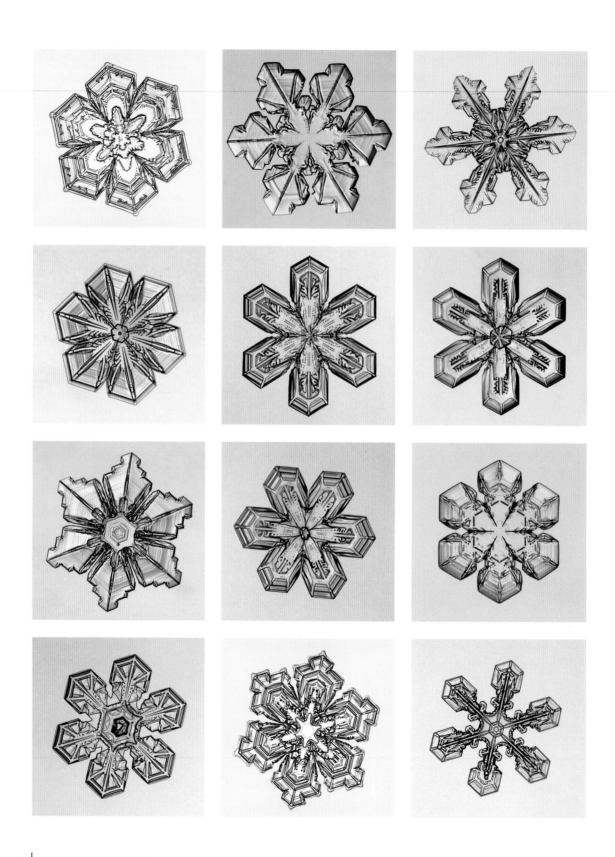

plate provides a good example of the resulting ambiguity. There are no absolute rules defining how prominent the ridges need be before a crystal is called a sectored plate. The dividing line between stellar plate and stellar dendrite is similarly a bit fuzzy. Naming snowflakes is not an exact science.

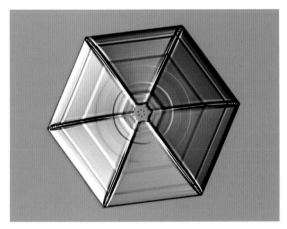

SIMPLE SECTORED PLATE | The archetypal sectored plate is a simple hexagon divided by ridges into six sectors.

COLUMNS AND NEEDLES

Columns and needles are the oft-ignored members of the snow crystal family. We daresay you will not find any columnar or needle-like snowflakes depicted in holiday advertisements at your local shopping mall. In the real world, however, these crystals are common, and many snowfalls are made predominantly of ice columns and needles. They are generally smaller than stellar crystals, so they are easy to overlook on your sleeve. A magnifier gives a good view, but a microscope is necessary to see their finer details.

Columns and needles form most often when the temperature is near 21 degrees Fahrenheit (−6 °C), although stout columns might also be found when the temperature is bitter cold, below around −13 degrees Fahrenheit (−25 °C). Their overall shape is that of a slender hexagonal column, like a wooden pencil, but the faceting is frequently rounded and not apparent. Thicker columns often contain conical hollows on either end.

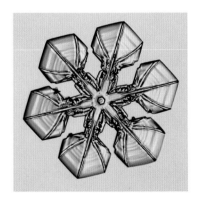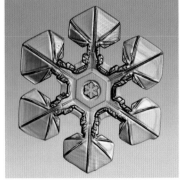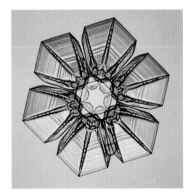

DUCK FEET | The partitioned branches of these three sectored-plate crystals resemble webbed feet.

EACH UNIQUE | Stellar plates (opposite) appear in a limitless assortment of shapes and patterns.

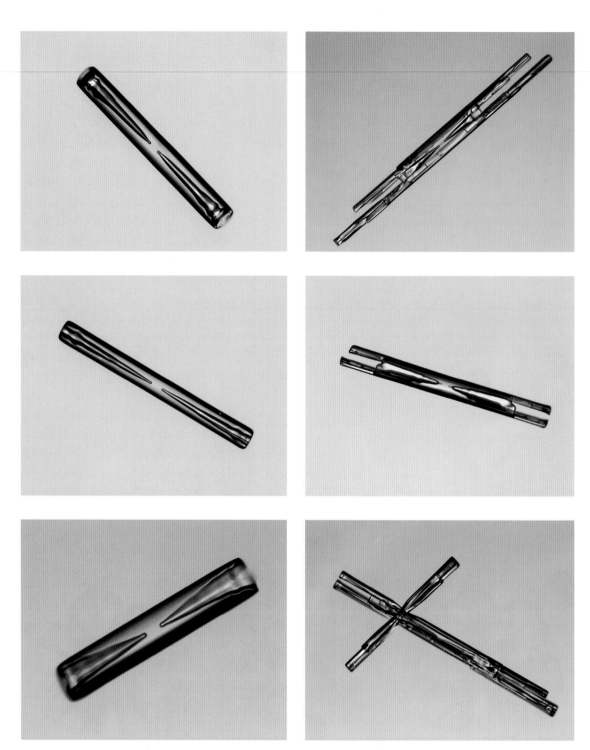

HOLLOW COLUMNS | Columnar crystals often develop with conical hollow spaces in both ends.

SPLIT ENDS | The ends of columns often split as they grow, yielding needle clusters.

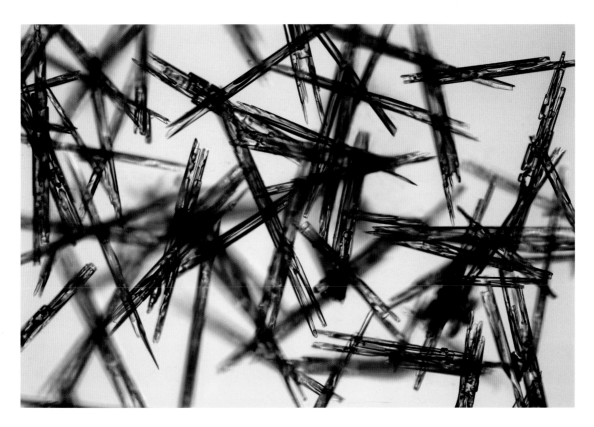

ON PINS AND NEEDLES | Some snowfalls bring mainly columns and needle crystals. We caught the ones in the photo above by setting a glass slide out in the falling snow for a few minutes.

BULLET ROSETTES | When multiple columns grow out from a common center, they are shaped a bit like bullets, and the cluster is called a "bullet rosette."

CAPPED COLUMNS

Capped columns have to be our favorite of the more exotic snow crystal types. First of all, they are common enough that you can find a few if you go looking for them. Second, they come in a wide variety of shapes and sizes, making them fun to look at. Finally, many of them are just plain odd-looking, and you know by now that we mean that in the most flattering way possible.

To understand the formation of a capped column, you just have to look at the morphology diagram and remember that the temperature can change while a crystal is growing. A typical capped-column scenario begins a bit below 21 degrees Fahrenheit (–6 °C), yielding a stout columnar crystal. Then the column moves to a colder spot, where the temperature is below 10 degrees Fahrenheit (–12 °C), and plates grow on each of its ends.

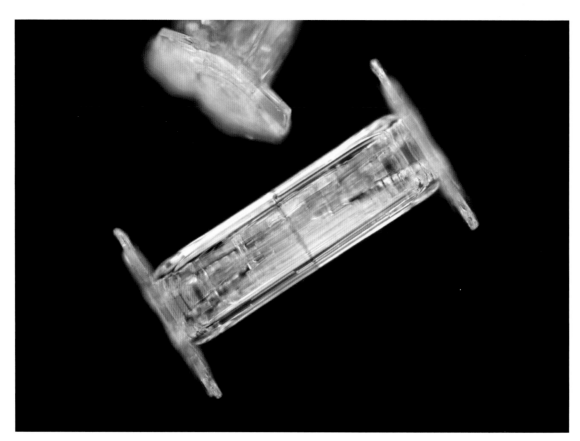

CAPPED COLUMN | This picture shows a classic capped column, as you might see it on your sleeve. We added a bit of red illumination when taking the photograph.

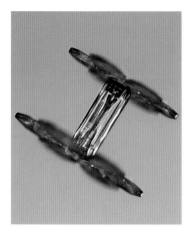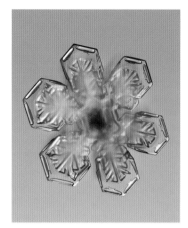

THREE VIEWS | These pictures show three views of a single capped column crystal: from the side (left); from one end, focusing on the top plate (middle); and the same view, but focusing on the bottom plate (right).

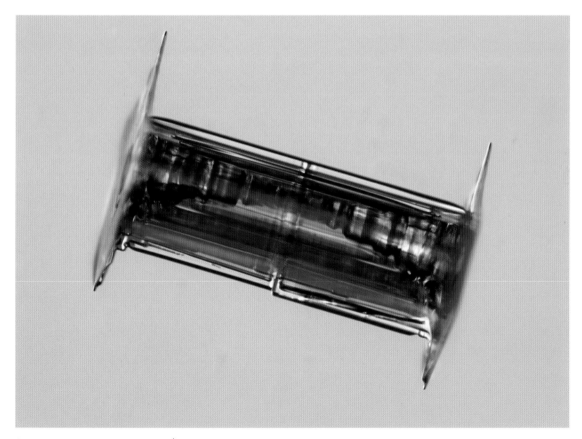

SURPRISINGLY COMMON | Capped columns have a distinctive shape and can be found in many snowfalls.

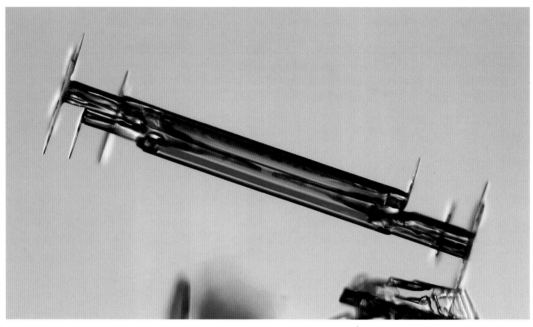

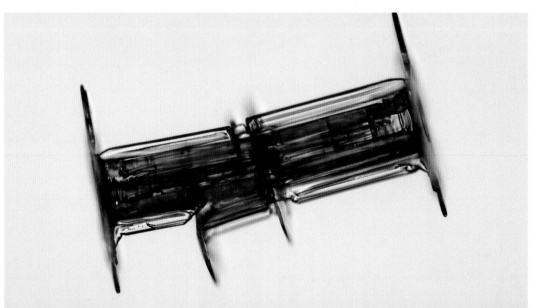

DOUBLE PLATES

A *double plate* is a short capped column where one plate is substantially larger than the other. When you look closely at stellar plates, you find that many are double plates. The smaller plate is like the sibling of a celebrity—it's there but often goes unnoticed.

The size discrepancy between the two plates arises because they begin growing in close proximity, so they compete for water vapor in the air. If one plate happens to stick out a bit farther than the other early on, then it begins to shadow its brother plate, hindering its growth. Soon one plate is almost completely shadowed by the other.

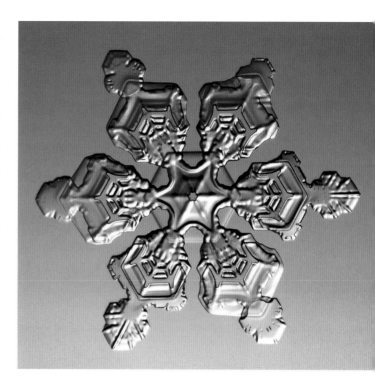

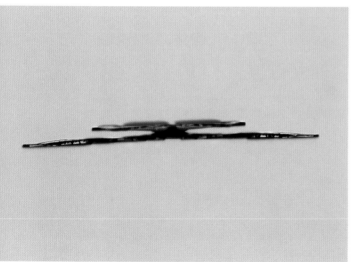

FLYING SAUCER | This photo shows a double-plate snow crystal seen from the side.

Opposite
MULTIPLY CAPPED COLUMNS | When columnar crystals develop two end plates and additional side plates, as shown in the examples on this page, they are called *multiply capped columns*.

HIDDEN HEXAGONS | At first glance, the crystal in the first photograph above appears to be a standard stellar plate. If you look closely, however, you can see a second plate; it is hexagonal in form, slightly out of focus, and about one-third the size of the main plate. The second picture shows something similar, this time with a distorted hexagonal plate. In each case, the two plates are held together by only the small nubs at the center of each crystal.

SPLIT PLATES AND STARS

Another variant of the capped column occurs when part of one plate grows to full size, accompanied by the complementary part of the other plate. We call these crystals *split plates* or *split stars*.

The formation of split crystals is similar to the double plates, providing another example of icy sibling rivalry. But instead of one plate overshadowing the other, different parts of the two plates become dominant.

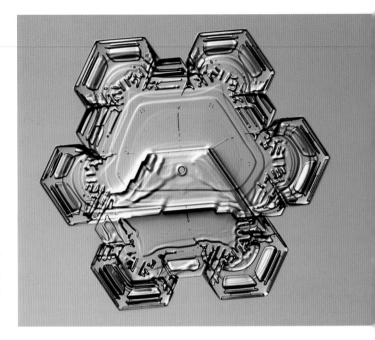

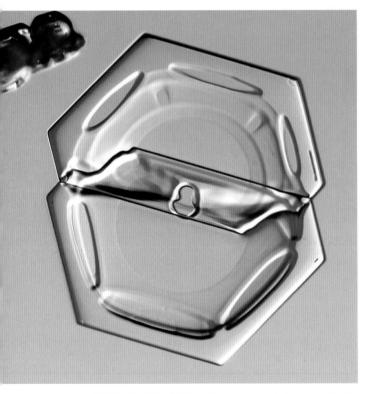

SPLIT PLATE | This split plate is made from two half-plates held together by a short central column.

SPLIT STARS | The split star in the top photo has two branches on one plate and four on the other. Only the tiny nub in the center holds the two parts together. The bottom photo shows one piece of a broken split star.

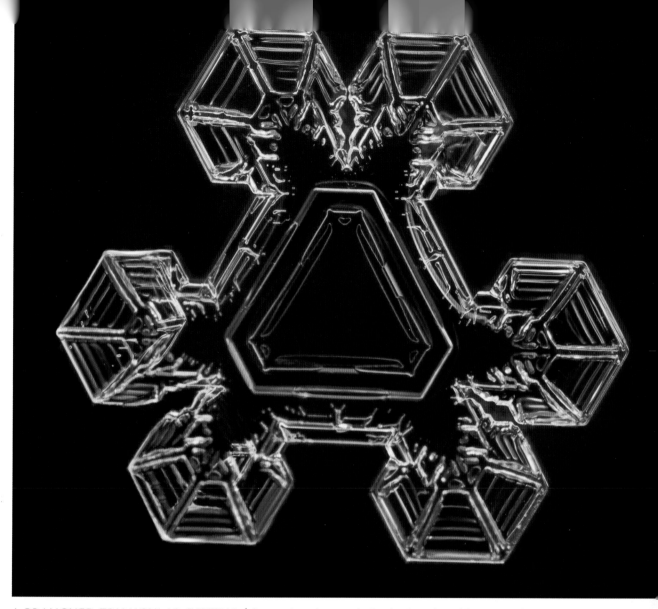

A BRANCHED TRIANGULAR CRYSTAL | Some triangular crystals develop branches with an unusual symmetry.

TRIANGULAR CRYSTALS

We are often surprised by how many *triangular crystals* we find. They are essentially the same as hexagonal plates, except three of the prism faces grew faster than the other three, yielding a triangle with truncated tips. How these crystals form is a bit of a mystery, but it appears that aerodynamic effects play a role in promoting the triangular shape. Triangular crystals are especially prevalent near 28 degrees Fahrenheit (–2 °C).

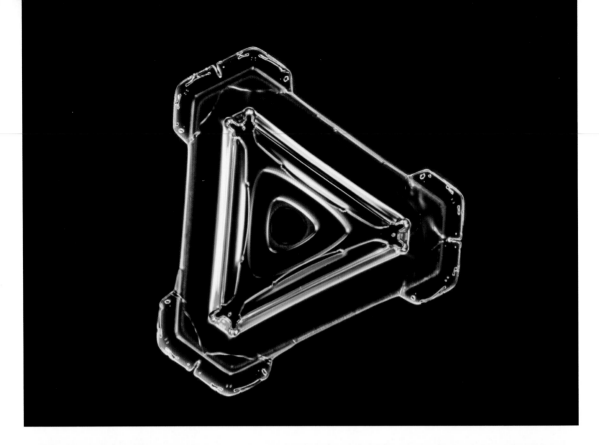

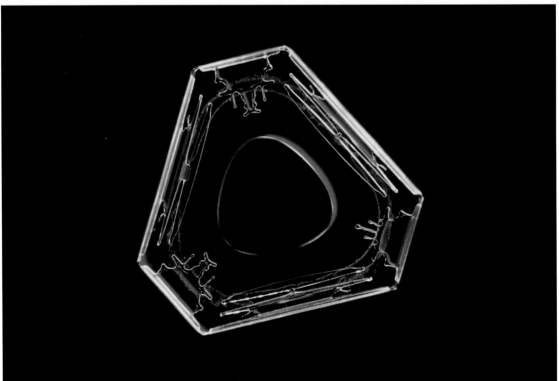

TRUNCATED TRIANGLES | Most triangular snow crystals are small plates.

TWELVE-BRANCHED STARS

Another surprisingly common snowflake type is the *twelve-branched star*, which looks like a normal stellar snowflake, except with twelve branches instead of six. It appears that two six-branched crystals collide in mid-flight to create this design. Because a twelve-branched star is not made from a single ice crystal, we call this a snowflake, not a snow crystal.

It remains a bit puzzling how random collisions produce so many nicely formed twelve-branched stars. The answer is probably just a selection bias when looking for photogenic specimens. If a collision goes well, with nearly coincident centers and a 30-degree separation between branches, then the snowflake grows into a twelve-branched star. If the collision doesn't go so well, the subsequent growth yields a messy jumble of ice. We tend to ignore the icy jumbles when photographing, while the well-formed stars catch our attention.

Regardless of their origin, these are distinctive looking snowflakes, and we are always pleased to find a good one. If you keep an eye out for twelve-branched stars, you will find them as well.

TWELVE-BRANCHED SNOWFLAKES | Each of these twelve-branched snowflakes is made from a pair of six-branched snow crystals joined together.

OBSCURED BY RIME | The six branches of this snowflake are covered by a thick coating of rime.

RIMED SNOWFLAKES

When a snow crystal collides with cloud droplets in flight, the droplets freeze on the crystal surface; in aggregate these droplets are called *rime*. If a snowflake is made mainly from rime particles, then it is called *graupel*, or *soft hail*. In our opinion, a dusting of rime is like a splattering of mud—it generally does not improve one's appearance.

BLEMISHED BEAUTY | This crystal grew to full size before accumulating a dusting of rime particles.

Frost crystals are related to snow crystals because both grow the same way—by absorbing water vapor from the surrounding air. Frost is not frozen dew, just as snowflakes are not frozen raindrops. Because they both grow from water vapor, frost crystals and snow crystals follow the same growth rules, yielding similar-looking structures. The snow crystal morphology diagram applies to both as well, so frost crystals appear plate-like or columnar at different temperatures.

When frost appears overnight on the grass, it usually forms minuscule ice crystals. The general appearance is like a dusting of white powder, and you can easily overlook the individual frost crystals. Nevertheless, if you get out your magnifier and get your face down there for a close look, you can see structures that resemble snow crystals.

Sometimes larger frost crystals grow on tree branches, fence posts, or other surfaces, and these are called *hoarfrost*. They can grow for extended periods to sizes that are many times larger than snowflakes, producing some gorgeous crystals. When hoarfrost grows near 5 degrees Fahrenheit (−15 °C), the

Don Komarechka

crystal structures are platelike and leafy, resembling large branches from a stellar dendrite snow crystal. When the temperature is near 21 degrees Fahrenheit (−6 °C), hoarfrost crystals form cups and scrolls that are essentially extra-large hollow-column snow crystals. Hoarfrost is not especially common unless a plentiful source of water vapor is nearby, for example from flowing water or a hot spring.

Hoarfrost often forms on the surfaces of snowbanks, and then it is called *surface hoar*. Being ice on ice, these crystals often go unnoticed. But when you see the sun sparkling off a snowbank, the bright sparkles are likely coming from the facets of surface hoar crystals that formed the night before.

For the case of *window frost*, the ice formations hug the glass surface and usually bear little resemblance to snow crystals. Window frost often follows minute scratches, streaks and swirls that were left behind when the glass was last cleaned. The intrinsic patterning on the glass surface combines with the ice crystalline structures to create a marvelous variety of displays. Your grandparents likely saw more window frost than you have, as it rarely forms on modern double-pane windows.

As with good snowflakes, beautiful frost crystals of all varieties are out there if you go looking for them. Our favorite hoarfrost sighting was many years ago on our family farm in North Dakota. Rachel lifted the seat cover in a long-abandoned outhouse and found some spectacular cup-shaped crystals on its lower surface. The cover was colder than the pit down below, so convection slowly brought water vapor up from the depths. In this secluded spot, the crystals had probably been growing undisturbed for weeks, and some of the cups were nearly the diameter of a finger.

BY ANY OTHER NAME

Just how many different types of snowflakes are there? We have examined the more common forms in this chapter, but in fact, there is no answer to this question. There is no absolute classification of different snowflake types, and there can never be one.

To understand the dilemma of naming snowflakes, consider a different question: how many named colors are there? There are the standards, of course—red, green, blue, yellow, orange, etc.—but is there a complete list of all color names? No, and there can never be a definitive list.

The names of colors depend on who is doing the naming. The makers of Crayola crayons, certainly a time-honored authority on the subject, inventory 133 standard colors, including Jazzberry Jam, Blizzard Blue, Outer Space, Neon Carrot, and Fuzzy Wuzzy. (We don't remember any of these from our crayon boxes growing up, but times have changed.)

The Sherwin-Williams paint company has its own table of more than 1,500 colors. The whites alone number 184, including Origami White, Feather White, Alluring White, Frosty White, Superwhite, Nouvelle White, Nebulous White, Nonchalant White, and Snowfall. You get the picture. While many color charts are being used for many purposes, there is simply no absolute list of color names.

We name snowflakes for the same reason we name anything—so we can more easily talk about them. Certain snow crystals are common and distinctive looking, and those have fairly well-defined names. Stellar plates, stellar dendrites, fernlike stellar dendrites, hollow columns, and capped columns have all been part of the snowflake vocabulary for some time. Go significantly beyond that, however, and opinions differ.

In an effort to be inclusive, perhaps, tables of snowflake types have become larger with time. In the 1940s the largest classification chart included 41 members. This number jumped to 80 in the 1960s, and recently a new table appeared with 121 different snowflake types.

The chart on the facing page shows our preferred snowflake classification. Because there can be no such thing as a final, definitive catalog, we pared the number down to make a chart that is convenient for snowflake watching. We describe many of these snowflake types in the preceding pages, and the chart gives you an idea of what other kinds of snowflakes you might encounter.

When you go outside to look at the falling snow, magnifier in hand, you can use this chart to guide your search for exotic snowflakes. You are more likely to spot a triangular crystal, a bullet rosette, or a double plate, if you know it exists.

Simple Prisms	Solid Columns	Sheaths	Scrolls on Plates	Triangular Forms
Hexagonal Plates	Hollow Columns	Cups	Columns on Plates	12-branched Stars
Stellar Plates	Bullet Rosettes	Capped Columns	Split Plates & Stars	Radiating Plates
Sectored Plates	Isolated Bullets	Multiply Capped Columns	Skeletal Forms	Radiating Dendrites
Simple Stars	Simple Needles	Capped Bullets	Twin Columns	Irregulars
Stellar Dendrites	Needle Clusters	Double Plates	Arrowhead Twins	Rimed
Fernlike Stellar Dendrites	Crossed Needles	Hollow Plates	Crossed Plates	Graupel

SNOWFLAKE FORMS | This chart depicts the most frequently observed types of snowflakes.

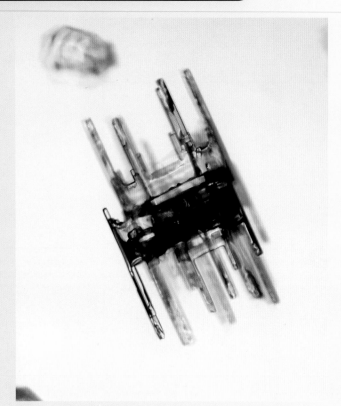

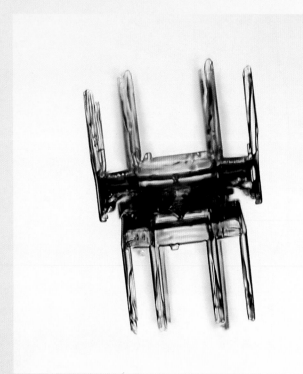

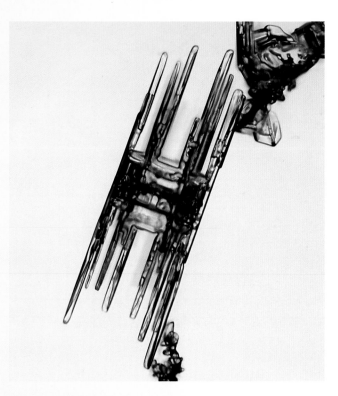

On one of our snowflake photography visits to Fairbanks, Alaska, we encountered the odd snow crystals pictured here. The weather was unseasonably warm that January day, with temperatures just a bit below freezing, and there was little wind. The falling crystals were small, so we spread out a half dozen glass slides as a collecting surface and then scanned the slides one by one under our photomicroscope to see what we could find.

Mostly we saw misshapen granular crystals that were not much to look at, but we kept searching. Then, all of a sudden, we began to find some truly bizarre crystals. From our chart of snowflake forms (page 61), you can see that the crystals shown here are a curious hybrid of several different varieties.

Using the morphology diagram, we can infer that these crystals experienced several abrupt temperature changes in the clouds. They must have switched from columnar growth to plate-like growth and then back to columnar again during their development. It is unusual for a single snow crystal to experience this many extreme temperature swings, so you will not encounter such odd crystals very often. The unusual atmospheric conditions that produced these crystals lasted only about ten minutes.

In sports, people say that you miss every shot you don't take. In snowflake watching, you miss all the crystals that fall when you're not looking. We run into offbeat crystals mainly because we just keep hunting. You never know what kinds of crazy snowflakes will appear from one minute to the next.

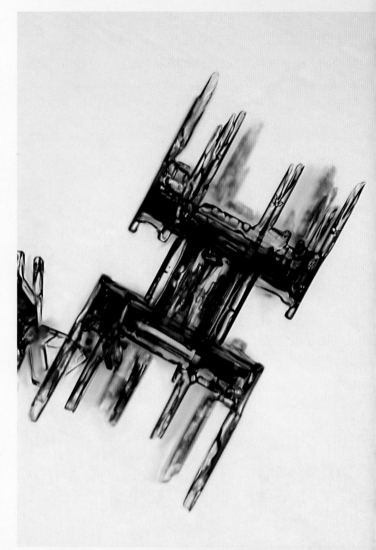

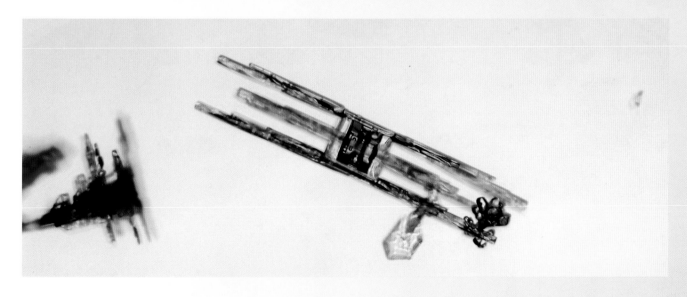

CHAPTER FOUR
Snowflake Weather

"Nature is ever at work building and pulling down, creating and destroying, keeping everything whirling and flowing, allowing no rest but in rhythmical motion, chasing everything in endless song out of one beautiful form into another."

—JOHN MUIR, *OUR NATIONAL PARKS*, 1901

Snowflakes are being manufactured in the atmosphere at an astounding rate—from snowfall data, we calculate around a million billion crystals each second. Every ten minutes, that's enough snow to make an unstoppable army of snowmen, one for every person in the world. Over the earth's history, some ten times the mass of the planet has floated down to its surface in the form of tiny ice crystals.

The nurseries that produce these vast numbers of snowflakes are none other than the clouds filling the sky on a winter's day. We can learn a thing or two about snowflakes by taking a look at the clouds in which they are created: where they come from, what they are made of, and how they bring forth snow crystals in such endless variety. Next time you are looking up at the gray winter sky and asking whether it will snow today, pause to consider what must be happening up there to generate these marvelous crystalline structures.

MAKING A SNOWSTORM

If you wanted to mimic nature and create a snowstorm, your first step would be to evaporate some water to make water vapor. A good-sized cloud bank contains maybe a million tons of water, give or take, so the pot on your stove is probably not sufficient for the task. In nature, this step is accomplished mainly through evaporation from oceans, lakes, and rivers, powered by heat from the sun.

When the water vapor is in the air, your next step is to cool the air down so clouds form. Cooling makes the relative humidity go up, and at 100 percent the air is said to be *saturated* with water vapor. Cool the air further and the humidity rises above 100 percent, and then the air is *supersaturated*. When this happens, the excess water vapor starts looking for a way to condense out.

If you supersaturate air close to the ground, say in your backyard, then the water vapor will condense out as dew on the grass. To produce

Opposite
BORN OF THE CLOUD | We found this stellar dendrite snow crystal in Cochrane, Ontario.

FREEZING A CLOUD

The next step in your homemade snowstorm is to continue cooling your parcel of air, now containing a cloud, until the liquid water droplets start to freeze. You might think this would happen right at 32 degrees Fahrenheit (0 °C), as that is the normal freezing point of water. But in fact you have to go colder, so the water droplets become *supercooled*.

The temperature needed to freeze a water droplet depends on what kind of dust it contains. Those captive dust particles *nucleate* the freezing process, and this can happen only below the freezing point. Moreover, some dust particles do this better than others. Typically, cloud droplets must be chilled to somewhere between 21 degrees Fahrenheit (–6 °C) and 5 degrees Fahrenheit (–15 °C) before they turn into ice. Remove the dust, and droplets of very pure water can be supercooled to nearly –40 degrees Fahrenheit (–40 °C) before they freeze.

A snowflake begins growing right after a microscopic cloud droplet freezes. As the speck of newly frozen ice floats through the cloud, it captures molecules of water vapor from the surrounding air and becomes larger. As a result, the air becomes slightly drier around the growing crystal, causing nearby liquid droplets to slowly evaporate away. It takes fifteen minutes to an hour to grow a good-sized snowflake. In this time about one hundred thousand nearby droplets will have evaporated away to supply the water vapor to make just one snowflake.

As a snowflake grows and becomes heavier, it eventually begins to fall. After it leaves the clouds, the snowflake no longer has a ready source of water vapor, so it stops growing. From then on the crystal drifts slowly downward with a typical velocity of around one mile per hour.

KITCHEN CLOUDS | Heating a pot of water puts water vapor into the air, which condenses into small droplets as the air cools right above the pot. As the droplets continue moving upward, they evaporate and disappear. You can see the cloud of liquid water droplets above the pot, but the water vapor is invisible.

clouds, you have to cool your parcel of air high above the ground. Up in the atmosphere, with no other surfaces available, the excess water vapor condenses onto microscopic dust particles that are always abundant in the air. The water vapor turns into tiny water droplets, and each one of these countless droplets contains a speck of dust. Although water vapor is an invisible gas, the water droplets are visible collectively as a cloud. Unlike raindrops, these cloud droplets are so small that they do not fall but remain suspended indefinitely as they are carried by the wind.

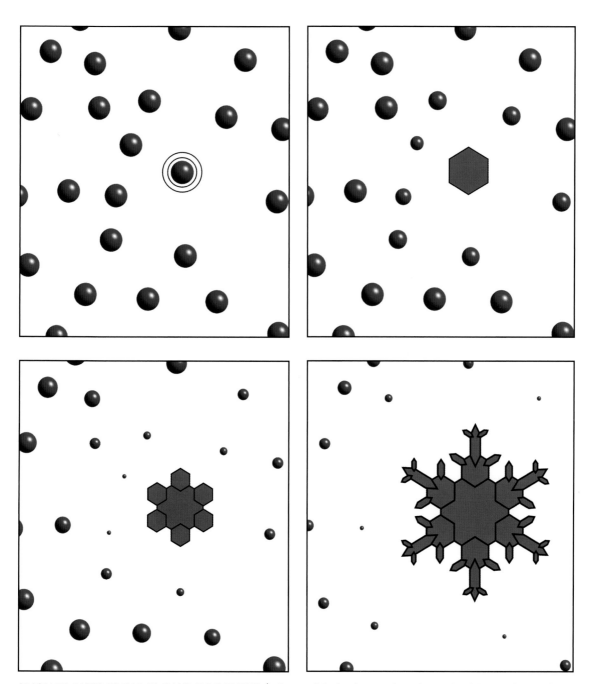

SNOWFLAKES FROM CLOUD DROPLETS | A snowflake begins growing when a cloud droplet freezes. The ice crystal grows by capturing water vapor from the surrounding air, causing nearby droplets to evaporate away. About one hundred thousand cloud droplets provide enough water vapor to make a single snowflake.

SUPERCOOLED CLOUDS | Winter clouds are often made of liquid water droplets that have been supercooled below the freezing point.

TOO COLD TO SNOW

When the weather is especially frigid, you may hear people say that "it's too cold to snow." Opinions vary as to when this old adage should be dusted off and exercised, but you might hear it any time the temperature is below 0 degrees Fahrenheit (−18 °C). You can see the truth in this statement by looking at the life of a snow cloud.

Clouds typically start out as liquid water droplets, even in winter. As the temperature falls, some droplets freeze and many more evaporate to make snowflakes, as just described. If the temperature continues falling, more droplets freeze and more snowflakes form. By the time the temperature has dropped to 0 degrees Fahrenheit (−18 °C), the remaining droplets are pretty much all frozen, and the cloud becomes a *cirrus cloud* made entirely of tiny ice crystals. With no more suspended liquid droplets, there is no excess water vapor for making snowflakes. Thus it has become too cold to snow.

There are few absolutes when talking about weather, so it is never absolutely too cold to snow. Even in Arctic climates, where the temperature might be a brisk −40 degrees Fahrenheit (−40 °C), you sometimes encounter a light flurry of sharply faceted diamond-dust crystals. But cold air is dry air, so growth rates are slow and the crystals are almost always tiny. It can snow at low temperatures, but it doesn't snow much.

GEOGRAPHY MATTERS

If you really did try to engineer your own snowfall following the previously mentioned process, you would find those cooling steps to be especially challenging. Nature gets the job done using a number of tricks, one of which is called *orographic lift*. Basically, the wind lifts the air by pushing it up the side of a mountain. The air expands as it goes to higher altitude, and this expansion cools the air. Push the air high enough, fast enough, and it becomes supersaturated. Clouds follow and then rain or snow.

The Sierra Nevada range in California sees a lot of orographic lift. The mountains intercept the moist winds coming from the Pacific Ocean, producing copious amounts of snow. The mountain town of Tamarack, California, is one of the snowiest places in North America as a result. One of Tamarack's precipitation records still stands from 1911, when the town experienced the most snow in the United States in a single month—a whopping 38 feet (11.5 meters).

A storm front is another trick nature uses for making snow. TV meteorologists are often pointing out the approaching storm fronts on weather maps, as these are where rain or snow is likely. In a nutshell, the distant winds push one air mass up and over another, producing a result that is much like orographic lift: the rising air expands and cools, yielding clouds and precipitation.

DOING YOUR PART

There is actually a bit of you in the snowflake pictures you see in this book. That's because even you, right this minute, are making a contribution to the atmospheric water supply. You are putting water into the air every time you exhale. In fact, you personally put so much water into the air that some of your water molecules almost certainly made it into these snowflakes.

You exhale roughly a liter of water per day, and most of that water rains or snows back down again somewhere within a week or two. The total global precipitation is around 1,000,000,000,000,000 times greater than the amount of water you exhale, so your impact on the weather is pretty minor. But even if you contribute only 1/1,000,000,000,000,000th of the total water content in a snowflake, that is still about 1,000 molecules. It depends on how well things are mixed in the atmosphere, but there are probably, very roughly, about a thousand of your molecules captured in every snowflake picture. Thank you for your contribution—and please, keep up the good work.

Andrey Arkusha/Shutterstock

LAKE-EFFECT SNOW | When cold air blows over the surface of a warmer lake, copious amounts of water evaporate into the air. Clouds form, and heavy snowfalls can result.

THE PERFECT (SNOW) STORM

We experienced one especially memorable winter storm front on January 5, 2006, in the small Canadian town of Cochrane, Ontario. It was a stationary front, so one invisible mountain of air hardly budged as another pushed slowly over it. That front stayed over us all day, yielding a continuous light snowfall. For photographing snowflakes, this was the perfect storm.

The temperature hovered right around 5 degrees Fahrenheit (–15 °C), perfect for growing large stellar crystals. The clouds were low, so the crystals landing on our collection board were all freshly made with sharp facets. There was little wind, and the most gorgeous crystals floated gently down for eight straight hours. We took more than 400 snowflake pictures that day, capturing some of the most beautiful stellar crystals we have seen. We were just about frozen stiff by the time that wonderful front finally moved on.

LAKE-EFFECT SNOW

Another one of nature's cloud-making tricks is to mix cold air with warm, moist air, yielding supersaturated air at some intermediate temperature. This happens when a cold wind blows over an unfrozen lake, and a good example is the north wind from Canada that brings frigid air over Lake Superior and onto the Upper Peninsula of Michigan. When the cold wind hits the warmer lake, you see steam rising up from the surface. Call it steam, call it fog, or call it a low cloud; these are all words for essentially the same thing—a mist of tiny water droplets. Carry that steamy air up and off the lake, and downwind you may experience a heavy snowfall. This phenomenon is called *lake-effect snow*.

Ken had a delightful stay one January in Houghton, Michigan, which sits near the south shore of Lake Superior and is thus well situated for lake-effect snow. The average snowfall in Houghton in January is about an inch per day, and that is what he experienced—about an inch, slow and steady, almost every day. The snow in Houghton is a lot like the rain in Seattle; it seems to be on more often than off.

For photographing snowflakes, all this snow was terrific, but the residents understandably found it wearisome. From his motel window, Ken watched one hearty fellow shoveling his driveway almost every evening. Maybe his snowplow was broken, or maybe he wanted the exercise, but he accomplished this task the old-fashioned way, with shovel in hand.

Then one day the near-constant snowfall let up and the sun came out. It was beautiful, as the landscape was covered with bright, new, sparkling snow. The fellow with the shovel used this fair-weather opportunity to remove snow from the roof of his garage, to avoid it collapsing from the accumulated weight. He shoveled all that snow off the roof and onto his driveway, which then had to be shoveled yet again.

ARTIFICIAL SNOW

Although Mother Nature has her own techniques for making snow, such as orographic lift and lake-effect snow, engineers have come up with a few tricks of their own. Ski resorts use these when nature is not keeping the slopes covered, and backyard snowmaking has even become a do-it-yourself project. As it turns out, for a few hundred dollars you really can make your very own backyard snowstorm—more of a sleet-storm, actually, made of frozen water droplets instead of snowflakes.

The standard method for making artificial snow is a small-scale version of the expanding-air trick. Take compressed air from a commercial

ARTIFICIAL SNOW | Compressed air expands and cools in this snow gun, quickly freezing water droplets in the process (above). *Matti Paavola*

Opposite
FROZEN DROPLETS | The resulting artificial snow consists of small sleet particles that do not show the ornate structures or symmetry seen in natural snowflakes.

air compressor, mix it with water in a pipe, and shoot the mix out through a nozzle. Voilà—instant snow. Step-by-step instructions can even be found online.

The compressed air expands when it comes out of the nozzle, and the expansion causes it to cool. At the same time, the nozzle makes a mist of fine water droplets, just like you get from a spray bottle in your kitchen. The expanding air cools the droplets, causing them to freeze in a split second, and out comes a spray of artificial snow. If you take a close look at the "snowflakes" in artificial snow (we have), don't expect any beautiful stellar crystals. The rapid cooling means the droplets simply freeze into small balls of ice.

So, is artificial snow as good as real snow? Artificial snow packs more densely than the natural stuff, because the particles are more compact than natural snowflakes. For basic downhill skiing and snowboarding, artificial snow is preferable to having bare patches on the slopes. But if you want to go powder skiing through an airy layer of frozen fluff, then you have to seek out the natural stuff.

The art of snowmaking is now so advanced that you can cover your lawn with white anytime and anywhere, even in summer. Compressed air doesn't have the cooling power to make summertime snow, but liquid nitrogen freezes those droplets with aplomb. This ends up being some rather expensive snow, but department stores in Southern California occasionally employ the Hollywood snowmakers to fill a parking lot with snow at Christmastime, as a treat for the kids. Shorts weather but with snow—there's a SoCal experience if ever there was one.

SNOW INDUCERS

Another method in the snowmaking business is to use chemical *snow inducers*. In a snow gun, water droplets must freeze during that split second they are in the nozzle; otherwise liquid water hits the slopes. Providing all that cooling requires a lot of compressed air, and running those air compressors nonstop can generate a hefty electric bill for the ski resort.

The role of the snow inducer is to cause the droplets to freeze more readily. Just as the dust particles in cloud droplets help them freeze, snow inducers increase the droplet freezing temperature. This means a lower electric bill, plus it allows snowmaking in warmer weather, which puts more snow on the slopes.

The hands-down favorite snow inducer comes from the bacterium *Pseudomonas syringae*. This little beasty produces proteins that nucleate freezing at 28 degrees Fahrenheit (–2 °C), holding the record for the highest ice nucleation temperature, higher than dust or any other material.

It appears that *Pseudomonas syringae* acquired its record-breaking ice-nucleation skills in order to inflict frost damage on plants. The bacterium is a plant pathogen, and damaging its hosts makes them more susceptible to bacterial infection. To improve its lot in life, *Pseudomonas syringae* therefore adapted and honed its ice-nucleation abilities.

We especially like the backstory that comes with *Pseudomonas syringae*. Scientists were studying the bacterium to better understand and mitigate frost damage on crops. But instead, some clever person realized that this microbe's talents could be harnessed for making artificial snow. The end result is better skiing at a lower price. You never know where science will take you.

LUNAR HALO | Light reflects off ice crystals high above the Earth to create this circular halo around the full moon. A 22-degree halo is occasionally visible around a bright moon. *Manuel Suárez Izquierdo*

ATMOSPHERIC OPTICS

When the sky is laced with thin cirrus clouds, you may be able to observe the sparkle of countless ice crystals in a display called an *atmospheric halo*. If the cirrus particles are predominantly roundish in shape, which is typical, then the cloud has the usual white appearance and there is not much else to see. But if the icy cloud contains well-formed hexagonal prisms—diamond-dust snow crystals—then light is deflected in certain preferred directions that are determined by the prism geometry. In these special circumstances, the combined effect of millions of individual sparkles from the faceted crystals produces an atmospheric halo.

The simplest halo is a circle of light that occasionally surrounds the sun or moon, with a 22-degree angle between the circle center and any point on the halo. If you close one eye, open your hand wide and place your thumb over the sun with an outstretched arm, then the tip of your pinky is about where you would find the 22-degree halo. (This rule of thumb—here involving a real thumb—works pretty well for adults and children alike, because arms and hands mostly scale with height.)

Although a suspended ice prism can deflect sunlight by any angle, a 22-degree deflection is especially likely. This statement follows from the geometry of the hexagonal prism, but it is not obvious without slogging through a lot of math.

The brightness of the 22-degree halo depends on the quality of the crystals. The most pronounced halos occur when sharply faceted hexagonal prisms are present. Less faceted ice crystals deflect light more randomly and do not produce bright halos.

The 22-degree halo is fairly common in winter, and it can be seen with some regularity wherever the climate is cold enough to require ownership of a snow shovel. The lunar halo is generally easier to see and is often noticeable even outside of snow country. Watch for it whenever the moon is full. Halo displays are more common than you might think; they often go unnoticed simply because people are not looking up.

Sun dogs, also known as 22-degree parhelia, are related to the 22-degree halo. Sun dogs are diffuse spots of light on either side of the sun that can appear when the sun is low in the sky. Sun dogs are visible when the suspended ice crystals have their basal faces oriented horizontally, which happens when falling crystals are aligned by the aerodynamics of their motion. Here again, sun dogs are fairly common in winter, although bright displays are rare.

SUN DOGS | This scene from Fargo, North Dakota, shows two bright sundogs on either side of the sun. *Gopherboy6956*

OTHERWORLDLY SNOW

Does it ever snow on other worlds? Possibly, but the snowflakes might look quite different from those found on Earth. On Mars, for example, water ice and carbon-dioxide ice (commonly known as dry ice) have both been spotted, and the latter can be several meters thick at the poles. It is still not known, however, if dry ice falls from the Martian atmosphere as "snowflakes" or forms directly on the surface like frost.

Even more peculiar is Titan, the largest moon of Saturn, where you might find methane snow. Methane is a gas at Earth temperatures, but becomes liquid on Titan, forming methane lakes and rivers. There appears to be a methane weather cycle on Titan that bears some resemblance to the water cycle on Earth, albeit at much lower temperatures.

Cirrus clouds of frozen particles appear on other moons and planets, including Titan, Mars, Jupiter, Saturn, Uranus, and possibly Neptune. Besides water ice, these extraterrestrial cirrus clouds may be made from dry ice, ammonia ice, or methane ice. Some of the clouds might even produce exotic atmospheric halos, because the different ices have different crystal symmetries compared to water ice. So far, however, the only planet definitively known to have falling snow and atmospheric halos is Earth.

MARTIAN ICE | This false-color photo of the surface of Mars shows water ice on the rocks and soil. It was taken by NASA's Viking Lander 2 at its Utopia Planitia landing site in 1979. The coating of frost is thinner than it looks in the picture; it's only about one-tenth as thick as a sheet of paper.

MARTIAN HALOS | This computer simulation shows what an atmospheric halo display might look like in the presence of octahedral and cuboctahedral dry-ice (carbon dioxide) crystals along with water-ice crystals. To date, atmospheric halos have only been observed on Earth, but the search goes on. *Les Cowley, www.atoptics.co.uk*

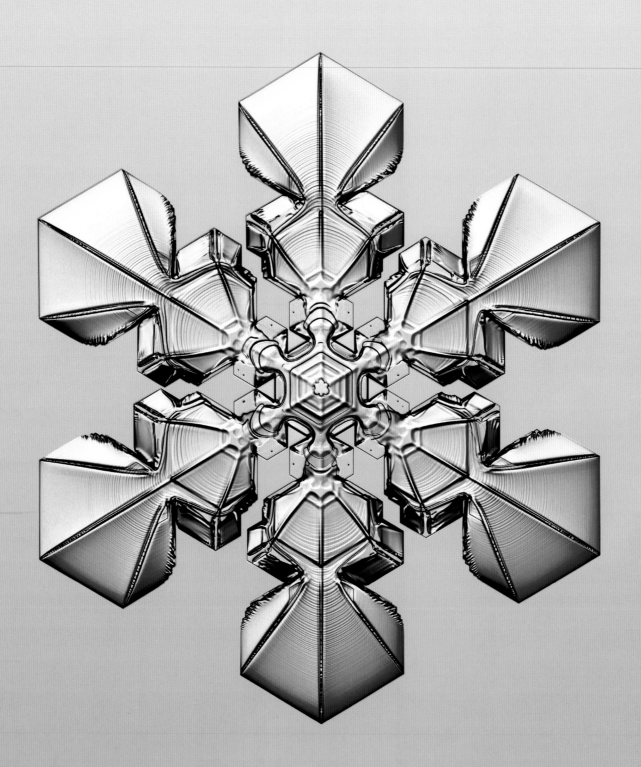

CHAPTER FIVE
Snow Crystal Symmetry

"The chief forms of beauty are order and symmetry and definiteness, which the mathematical sciences demonstrate in a special degree."
—ARISTOTLE, *METAPHYSICS*, CA. 330 BC

Symmetry abounds in nature. Animals that need balance and agility for running, like us, often show bilateral, or left/right, symmetry. Flowers display many different types of radial symmetry, as we see when four, five, or six petals are arranged equally around a common center. Fivefold symmetry is found in starfish and sea urchins, and if you cut an apple in half crosswise to separate the bottom from the top, you will see five seeds arrayed around the core like a star. And, of course, one of nature's most recognizable symmetries is that of the six-pointed snowflake.

The shape of a snow crystal is so distinctive that it almost defines its symmetry class—when you see six identical branches growing out from a common center and each branch exhibits bilateral symmetry, then you might well describe that object as being "snowflake shaped." It requires only a sheet of paper and a pair of scissors to depict, and the origin of this characteristic form can be explained by looking at how water molecules connect together to form an ice crystal.

SIX PETALS | Lilies often display six petals, giving them a symmetry that is similar to snowflakes. *a454/Shutterstock*

Opposite
CRYSTAL FACETS | Many snowflakes exhibit faceted features that reflect the hexagonal symmetry of the ice crystal.

WHY SIX?

The first scientist to theorize about the six-fold symmetry of snow crystals was German scientist Johannes Kepler. In 1611, Kepler presented a small treatise entitled *The Six-Cornered Snowflake* to his patron, Holy Roman Emperor Rudolf II, as a New Year's Day gift. In his treatise, Kepler contrasted the six-fold symmetry of snowflakes with similar symmetries found in flowers. He deduced that the similarities must be in appearance only, because flowers are alive and snowflakes are not:

> *Each single plant has a single animating principle of its own, since each instance of a plant exists separately, and there is no cause to wonder that each should be equipped with its own peculiar shape. But to imagine an individual soul for each and any starlet of snow is utterly absurd, and therefore the shapes of snowflakes are by no means to be deduced from the operation of soul in the same way as with plants.*

Kepler saw that a snowflake is really a relatively simple thing, made only from ice, compared to the baffling complexity of life that is embodied in a flower. He believed it could be fruitful, therefore, to question what organizing principle was responsible for snow crystal symmetry. Because it was known that cannonballs display a hexagonal pattern when stacked in a pile, Kepler conjectured that these two symmetries might be related. There was a germ of truth in this reasoning, because the geometry of stacking atoms lies at the heart of snow crystal symmetry. But the atomistic view of matter had not been developed by Kepler's time, so he could not carry the cannonball analogy very far.

Kepler realized that the genesis of crystalline symmetry was a worthy scientific question. He also recognized the similarity between snow crystals and mineral crystals, both exhibiting symmetrical faceted structures. But, at the end of his treatise, Kepler accepted that the science of his day was not advanced enough to explain it. He was certainly

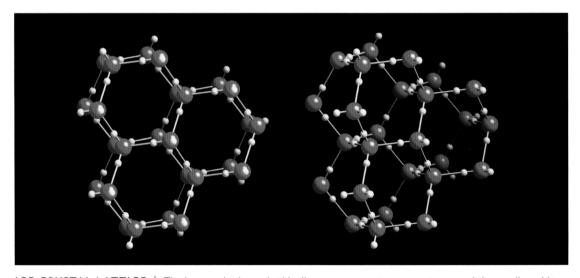

ICE CRYSTAL LATTICE | The large red spheres in this diagram represent oxygen atoms, and the smaller white ones represent hydrogen atoms. Each oxygen is flanked closely by two hydrogens, making an H_2O molecule. If you look at the crystal face-on (left), you can see the hexagons in the ice crystal structure.

EXAMINING SYMMETRY | German scientist Johannes Kepler made numerous contributions to astronomy, mathematics, and optics, including his famous discovery of what are now called Kepler's Laws, describing the motions of the planets around the sun. Kepler also took an interest in the precise six-fold symmetry of snow crystals, and sought to understand its origin. *Unknown artist, 1610*

HEXAGONS APLENTY | How many hexagons can you spot in these snowflakes? Hexagonal patterns are especially common on smaller, simpler snow crystals like these. Some are like tree rings, indicating times when the growth of the crystal changed abruptly.

correct in this conclusion, for three centuries would pass before scientists knew enough about atoms, molecules, and their arrangement in solid materials to finally answer Kepler's 1611 query.

The ultimate origin of snowflake symmetry lies in the structure of the ice crystal, specifically the regular arrangement of its water molecules. Each molecule is composed of one oxygen atom flanked by two hydrogen atoms, making the familiar H_2O. The three atoms bind together in a dog-leg-shaped molecule, and these in turn link up into sheets of slightly warped hexagons. Hexagons have six-fold symmetry, and this symmetry carries over into snow crystals. The precise geometry of the water molecule is what determines the structure of the ice crystal, and this in turn leads to the six-fold symmetry of a snowflake.

CRYSTALS

The word crystal derives from the Ancient Greek *krystallos*, meaning "ice" or "rock ice." Contrary to what the definition implies, *krystallos* was not originally used to describe ice but rather the mineral quartz. The early Roman naturalist Pliny the Elder described clear quartz *krystallos* as a form of ice, frozen so hard that it could not melt. Pliny was quite mistaken on this point; quartz is not a form of ice, nor is it even made of water. Nevertheless, after nearly 2,000 years, Pliny's misunderstanding is still felt in the language of the present day. If you look in your dictionary, you may find that one of the first definitions for crystal is simply "quartz." This is like saying the definition of food is "potato." Funny how some quirks in the language remain after thousands of years.

PARALLEL BRANCHES | A snow crystal is a single crystal of ice, even when it has a complex shape. The water molecules are all lined up in a rigid lattice, from tip to tip. This underlying hexagonal framework guides the growth of the branches, ensuring that sidebranches on one side of the crystal are parallel to sidebranches on the opposite side.

When we say crystal, we usually mean the scientific definition—a material in which the atoms or molecules are lined up in a rigid lattice structure. In addition to ice, all sorts of crystals can be found in our everyday lives. Copper is crystalline, as are ruby and diamond. Computer chips are made from silicon crystals. Most rocks are made from jumbled bits of crystalline minerals, such as quartz and feldspar. Salt, sugar, and aluminum foil are a few crystalline materials you can pick up at your grocery store.

Crystals can have other symmetries besides hexagonal. In ordinary table salt, for example, the atoms line up to form a cubic structure. If you take a close look at the salt on your dinner table (best done using a magnifier), you can see that the individual salt grains are mostly tiny cubes.

The symmetry of a crystal comes from how its constituent molecules stack together, and stacking things is a richer subject than one initially thinks. In the crystal world, there are thirty-two possible ways to stack molecules, including five different cubic forms and seven different hexagonal ones. Some symmetries are forbidden—there are no crystal structures with five-fold symmetry, for example. This is true for the same reason you cannot tile your floor with pentagonal tiles; pentagons simply do not fit together without leaving gaps.

Even spherical cannonballs can be stacked in different ways, into either hexagonal or cubic structures. In the warships of old, sailors stacked in pyramids to better distribute the weight of the cannonballs, and both square-bottomed and triangle-bottomed pyramids were used. Some of the finest mathematical minds in history have pondered the subtler aspects of how spheres can be stacked together, known in math circles as the *cannonball problem*. Itemizing and characterizing all the possible ways things can be stacked together

SALT CUBES | Different crystals have different symmetries. Grains of table salt look like tiny cubes, reflecting the cubic structure of the salt crystal lattice.

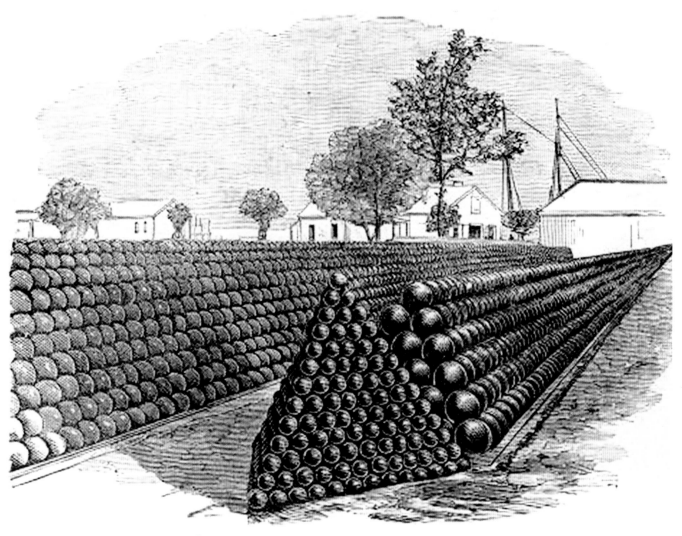

STACKING GEOMETRY | The stacking of molecules is governed by geometry, as is the stacking of cannonballs. The latter can be arranged into pyramids with either square or triangular bases. *Stacy*, Harpers Weekly

turns out to be a remarkably difficult undertaking. Arranging those oranges at the supermarket is perhaps not as simple as you thought.

Ordinary ice crystallizes into one specific hexagonal lattice structure and not any of the other thirty-one possibilities, because of how water molecules like to hook together under normal environmental conditions. At more extreme temperatures and pressures, especially very high pressures, water solidifies into sixteen additional crystalline phases. One of these, a cubic form of ice, is only slightly less stable than the hexagonal form we know so well. Had that dog-leg bend in the water molecule been just a few degrees different, you might be reading now about a completely different kind of snowflake.

CRYSTAL CONFUSION

If you immediately think of glassware when you hear the word *crystal*, you are not alone. It is one of those confusing terms, a bit like *flammable* and *inflammable* (which, amusingly enough, both mean *flammable*). As is often the case, these linguistic inconsistencies have their roots far back in history.

At the molecular level, glass is the very opposite of a crystal, although this was not known when the language originated. On one hand, there is the scientific definition of crystal—a material with a rigid lattice of atoms and molecules. Glass, on the other hand is an *amorphous* solid—its molecules are completely disordered, randomly jumbled atop one another. So a "crystal glass" is, at the scientific level anyway, a perfect oxymoron—it suggests a material that is inexplicably both ordered and disordered.

Oxymoron or not, the term "crystal glass" is likely in widespread use at your local department store. It is sometimes reserved for leaded glass, which is about one-quarter lead-oxide by weight. Leaded glass is heavier than ordinary glass, with a brighter sparkle, and it has a more satisfying ring when you strike it. Leaded glass is also not entirely safe for household use (because of the lead, as you might have guessed), so it is not as common as it once was.

Different metal oxides are often used in place of lead, and the resulting glass may also be called crystal glass. Of course some stores simply co-opt the term to refer to any glass items they can get away with charging you more for. Consumer protection agencies are beginning to regulate what types of glass can be called "crystal" for that reason. In the European Union, for example, "lead crystal" must be composed of at least 24 percent lead oxide, while "crystal glass" must include similar amounts of other metal oxides.

This bipolar use of the word crystal appears to go back centuries, long before anyone knew about even the existence of atoms and molecules, let alone how they are arranged in different materials. Glassmakers produced beautifully faceted *objets d'art* that resembled mineral crystals, and somewhere along the road people adopted *krystallos* for both. Whatever the origin, we are stuck with these dual meanings of the word crystal, in spite of any confusion it might generate.

THE LITTLE FACES

Although the molecular latticework of the ice crystal ultimately explains the six-fold symmetry of a snowflake, this is not the whole story. We now need to delve into how forces operating at the molecular scale guide the growth and development of a large snow crystal, and for this we need to look into snow crystal facets, along with how those facets arise.

The flat basal and prism surfaces of a hexagonal prism define the principal ice-crystal facets. The word *facet* literally means "little face," and the sparkles you see when the sun shines on a snowbank are reflections from the countless little faces of ice crystals. You can also see the mirror-like surfaces of individual snow crystals

when you look at them on your sleeve. As you move a crystal to and fro, bright reflections reveal the smooth facet surfaces.

This aspect of snow crystal construction was first recorded by René Descartes, who described snow crystals in 1637 as "little plates of ice, very flat, very polished, very transparent, about the thickness of a sheet of rather thick paper . . . but so perfectly formed in hexagons, and of which the six sides were so straight, and the six angles so equal, that it is impossible for men to make anything so exact." Descartes was describing snow crystal facets, and he pondered the origin of these precisely formed, mirror-like surfaces. Where do faceted surfaces come from? Why are they flat, and how do they form?

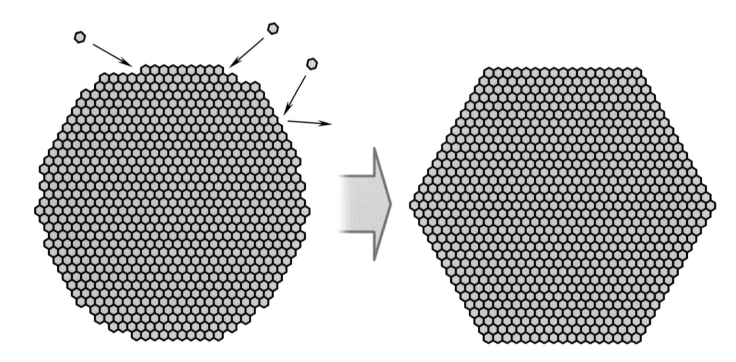

FACET FORMATION | Facets appear as a snow crystal grows. Molecules fill in rough areas on the surface, leaving only the smooth facet surfaces. In ice, these surfaces define a hexagonal prism.

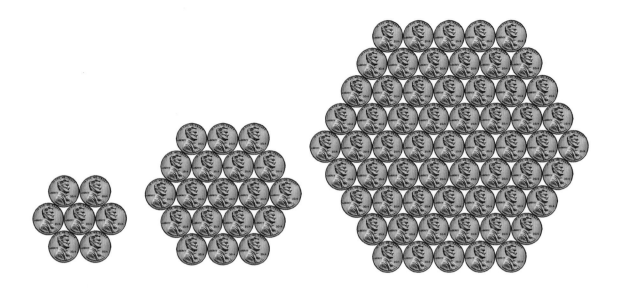

PENNY CRYSTAL | Arranging pennies using simple rules creates a hexagonal penny "crystal." If each penny represents one water molecule, then a small snow-crystal hexagon would cover the Hawaiian island of Oahu, and you would need $100 billion in pennies to make just one layer of molecules. That is one large jar of pennies.

SELF-ASSEMBLED FACETS

Consider a small, round crystal of ice as it begins growing into a snowflake. Water molecules from the air strike the surface of the crystal and attach, but they prefer to attach to the rough spots on the surface. Smooth surfaces are difficult to hold onto, while the rough spots have lots of dangling molecular bonds to grab. As a result, the rough spots accumulate molecules quickly, while the smooth surfaces do so more slowly. Before long, the rough areas add water molecules and fill in, leaving only the smooth areas to define the shape of the crystal. These smooth, slower-moving surfaces become the crystal facets.

Although you cannot see molecular faceting directly, you can simulate it using nothing more than a jar of pennies. Start by placing two pennies side by side on a table. This is your seed crystal, which you will grow by adding more pennies. Each time you add a penny, place it so it touches two or three other pennies. If a three-penny spot is available, use that first. If all you have are two-penny spots, pick one at random. Keep doing this, following those simple rules as you lay down pennies, and you will soon end up with a hexagonal penny "crystal." (Perhaps we can call it a *penny-gon*.) The more pennies you place, the more hexagonal your penny crystal becomes.

These simple rules for putting down pennies do not dictate any large-scale structure directly. All any penny knows is that it prefers three neighbors to two. Water molecules are similarly self-serving. All any molecule knows is that it would like to connect to as many other molecules as it can. So each molecule minds its own business, maximizes its own lot, and the end result is a faceted hexagonal prism.

HEXAGONAL ARCHITECTURE | Frank Lloyd Wright experimented with hexagonal geometry in several of his buildings. The Wall House in Plymouth, Michigan, is called the "Snowflake House" because of its hexagonal floorplan. *Map data © 2015 Google*

The process of faceting during crystal growth is one of the most important elements needed to explain the shapes of snowflakes. Faceting is why the large-scale structure of a hexagonal prism reflects the small-scale structure of the underlying molecular lattice. It is this process of faceting that allows the geometry of a water molecule to influence the geometry of a snowflake.

ARTIFICIAL FACETS

We should add that not all facets one sees arise naturally during crystal growth. The facets in gemstones, for example, are almost always artificial, regardless of the type of stone. The diamonds, emeralds, and rubies you find in jewelry all sparkle from facets that were carved and polished with a grindstone. The facets on crystal glassware were also carved or molded into place by the glassworker. These faceted works of art can be stunning, to be sure, but they do not possess the natural beauty of the self-faceting snowflake.

No grindstone is needed to make natural facets. They arise spontaneously during the growth of crystalline materials. Quartz crystals, such as the purple-hued amethyst, can simply be pulled out of the earth in beautiful faceted shapes, as can garnets and many other gemlike minerals. Sugar can be grown into large faceted crystals as well, often sold on a stick as "rock candy." You can grow beautiful crystals from alum (sold in powder form at your local grocery store) right in your own kitchen, and you can purchase crystal-growing kits that include a variety of self-faceting crystalline materials.

While artificial facets can have any chosen shape or orientation, natural facets are tied to the molecular geometry of the crystal. Natural crystal facets always display a symmetry that reflects the underlying molecular lattice.

THE SIX-FOLD WAY

We live in a largely rectilinear world—our street grids, houses, and computer screens are nearly all based on right angles. How refreshing it can be to encounter a different architecture, especially one expressed in something as beautiful as a snowflake.

The geometry of ice eschews the unremitting use of 90-degree angles. Instead, together with the lily and the honeycomb, snow crystals express 60- and 120-degree angles in their design motif.

As snowflake aficionados, one of our pet peeves is the all-too-common appearance of four- or eight-sided snowflakes in holiday advertisements. In part, this reflects our scientific side, wanting to see the proper snowflake symmetry being used. But beyond that, we feel that alternate symmetries present something of an affront to nature, converting the naturally beautiful six-fold architecture of the snowflake into one that better fits our regimented, rectilinear world.

Whenever possible, we try to nudge popular snowflake culture in the right direction. Recently Ken applied his unusual expertise during a brief side job as the snowflake consultant on Disney's animated movie *Frozen*. Basically he went over to the studio and described the structure and distinctive six-fold symmetry of snow crystals in detail, adding that real snowflakes are never four- or eight-sided. Thankfully, the artists at Disney adhered strictly to the proper ice crystal symmetry in all their animated snowflakes and ice palaces. If you watch the movie closely, you will not see an eight-sided snowflake anywhere. Did this make the movie better? Well, we like to think it did, in its own small way at least. That's our mission—increasing snowflake awareness.

THE RIGHT WAY TO MAKE A PAPER SNOWFLAKE

These photographs show how to fold a sheet of paper to make a stellar snow crystal with the proper six-fold symmetry. For you advanced paper-snowflake artists, we invite you to place your cuts to fashion anatomically correct snowflakes. In the snowflake shown here, for example, we included a hexagonal pattern near the crystal center, ridge-like slits centered on the main branches, and sidebranches at the proper 60-degree angle from the main branches. Features like these are all common in natural snowflakes.

To get started, consider it a challenge to construct paper snowflakes that model the shapes and surface patterns seen in the snowflake photographs on the facing page or perhaps elsewhere in the book. We warn you—it's not as easy as it sounds.

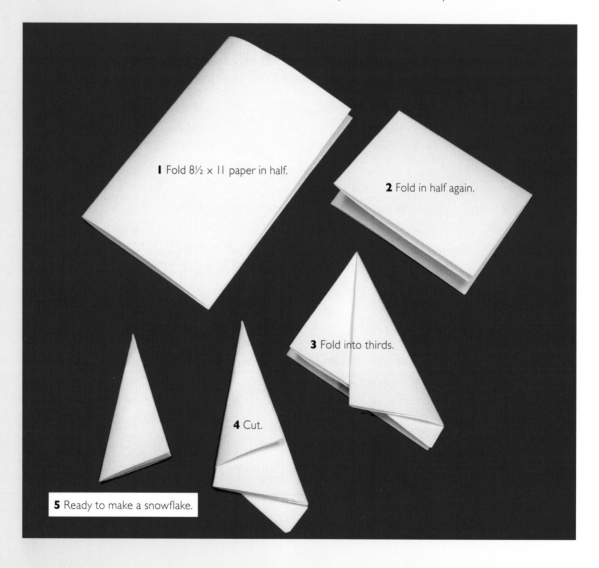

1 Fold 8½ × 11 paper in half.

2 Fold in half again.

3 Fold into thirds.

4 Cut.

5 Ready to make a snowflake.

CHAPTER SIX

Morphogenesis on Ice

*"The pursuit of truth and beauty is a sphere of activity
in which we are permitted to remain children all our lives."*
—ALBERT EINSTEIN, *LETTER TO ADRIANNA ENRIQUES*, 1921

The transformation that turns amorphous water vapor into an intricate stellar snowflake is an example of physical *morphogenesis*—the spontaneous creation of pattern and form. There are plenty of examples of physical morphogenesis you can find around you, such as ripples on ponds, banded patterns on snowdrifts and sand dunes, and even the billowing convection cells you see in a bowl of hot miso soup. The snowflake case is especially dramatic, as we see amazingly complex structures emerge, quite literally, out of thin air. The snowflake is the quintessential example of physical morphogenesis.

As mentioned in the previous chapter, faceting is an important player in the genesis of snow crystal structure. Faceting explains the formation of simple hexagonal prisms, defining the snow crystal's six-fold symmetry. But most snowflakes are much more elaborate than simple prisms, so faceting is only part of the story.

Our next task, therefore, is to explain why snow crystals grow into such complex, branched structures. Beyond that, we should also explain how the six branches manage to grow so similarly and why snow crystal patterns vary as much as they do. Although that sounds like a lot of explaining, the good news is that it all falls into place once we understand the origin of snowflake branching.

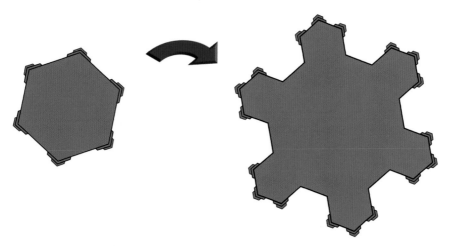

ORIGIN OF BRANCHING | The six corners of a snow crystal grow a bit faster because they stick out farther into the humid air, causing branches to sprout. As the crystal grows larger, the same effect causes sidebranches to sprout from the faceted corners of each branch. This process is responsible for the complex shapes of snow crystals.

Opposite
COMPLEX SYMMETRY | This stellar snow crystal displays aspects of both complexity and symmetry.

THE BRANCHING INSTABILITY

To begin, take a close look once more at the growth of a snowflake. As a crystal drifts slowly through the clouds, it captures water vapor from the air that surrounds it. The existing ice lattice assimilates water molecules and grows larger. As the crystal grows, however, it consumes excess water vapor around it, depleting the nearby air and reducing its humidity. For the crystal to keep growing, water molecules from farther away must diffuse through the air into the depleted region near the crystal, and this takes time.

The time it takes for water molecules to diffuse through the air is the underlying cause of snowflake branching. Here's how it works: Start with a simple, hexagonal crystal floating through a cloud. The six corners of the hexagon stick out a bit farther into the humid air, so they collect water vapor molecules a bit faster than anywhere else on the crystal. The corners thus grow a bit faster, and before long they stick out even farther than they did before. The corners experience runaway growth as this cycle accelerates—the corners stick out a bit, so they grow a bit faster; soon they stick out even more, so they grow faster still. Thus branches sprout from the six corners of a hexagonal prism.

This process of runaway growth is called the *branching instability*, and it is responsible for much of the elaborate structure you see in snowflakes, from hollow columns to stellar dendrites. The devil is in the details, and it takes sophisticated computer modeling to examine every sidebranch and faceted corner. But the essential process—driven by a runaway growth behavior—is one of the most important elements of snow crystal formation.

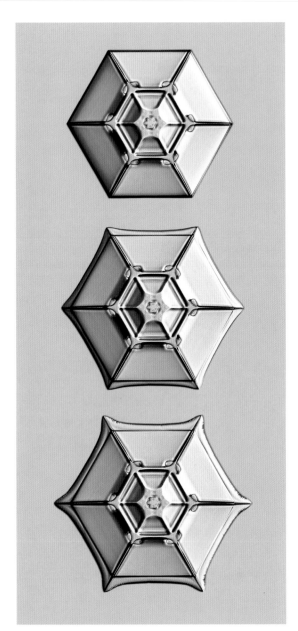

SPROUTING BRANCHES | This composite image shows three photographs of a small snow crystal as it grew in our laboratory. The crystal started out with a hexagonal outline (top), but then branches sprouted from the six corners.

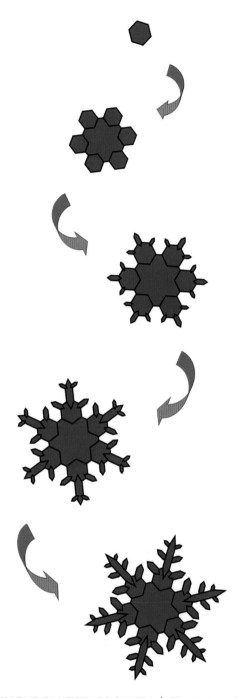

SYNCHRONIZED GROWTH

The pieces of the puzzle really fall into place when you bring faceting and branching together. For faceting alone yields only simple prisms, while branching alone tends toward complex but random structures. Acting in concert, however, these two forces create snow crystals that are both complex and symmetric.

Consider, if you will, the life story of an individual snow crystal, specifically a large, symmetrical snow star that you might catch on your mitten during a quiet snowfall. In the beginning, your crystal is born as a tiny nucleus of ice when a cloud droplet freezes. The frozen droplet initially develops facets as it grows, turning the nascent snowflake into a well-formed single crystal of ice—a minute hexagonal prism.

While still in its youth, fate places the crystal in a region of the clouds where the humidity is just right and the temperature is a perfect 5 degrees Fahrenheit (–15 °C). There the tiny crystal grows into a thin, flat, hexagonal plate. In this juvenile phase of its development, the crystal shape is being determined mainly by the process of faceting.

As it reaches snow crystal adolescence, the flake blows suddenly into a region of the clouds with higher humidity. The increase in water supply makes the crystal grow faster, which drives the branching instability. The initial hexagonal structure guides the formation of six branches, one on each corner of the hexagon. The branches grow independently of one another, but the higher humidity around the entire crystal causes all six branches to sprout at the same time.

The crystal subsequently blows to and fro, following the will of the wind. If it moves to a region of low humidity, faceting will add corners to the ends of the branches. If it moves back to higher humidity, sidebranches will sprout from the corners defined by the facets. Each change in its local environment alters the way the crystal grows, and each change is felt by all six branches simultaneously. Thus the six branches develop in synchrony while the crystal dances through the clouds.

As the crystal grows larger and more ornate, it eventually becomes so heavy that it floats gently downward, out of the clouds to land on your mitten. The exact shape of the crystal reflects the history of its growth. The six branches are nearly identical because they all experienced the same history.

People often think that the six branches of a snow crystal must somehow communicate with one another to coordinate their growth, but this is not the case. The six branches all grow independently of one another, but they grow alike because each experiences the same external fluctuations in temperature and humidity.

You can see an analogous coordination if you watch how people dress outdoors. On cooler days, you might see a lot of sweaters and scarves. On rainy days, raincoats appear. Did all these people talk among themselves to coordinate their dress? Of course not. Everyone simply chose outerwear that was appropriate for the weather; no communication was necessary. In the same way, the six branches of a snow crystal are synchronized by external conditions, not by internal communication.

SIMILAR SNOWFLAKES | These two snow crystals fell from the clouds within minutes of one another in northern Sweden. They look similar because they both followed similar paths during their formation. But their paths were not exactly identical, so neither are the crystals.

A GROWING SNOWFLAKE | This series of images shows a single laboratory-made snow crystal at different times during its growth. The humidity was changed as the crystal grew, so the growth behavior alternated between faceting and branching, yielding an especially symmetrical stellar snowflake. The initial hexagonal crystal was smaller than the diameter of a human hair, while the final stellar snowflake measured just over 2 mm (0.1 inches) from tip to tip. The total growth time was twenty-seven minutes.

NO TWO ALIKE

We've told the life story of an individual snow crystal, and now we can explain why snow-crystal patterns are so different from one another. The detailed morphology of each falling crystal is determined by the path it takes through the clouds and by the temperature and humidity it experiences along the way. A complex path yields a complex snowflake. Because no two crystals follow precisely the same path through the turbulent atmosphere, no two snowflakes will be exactly alike.

So the creative genius responsible for the endless variety of beautiful snowflake patterns lies simply in the ever-changing winds that guide each crystal's journey.

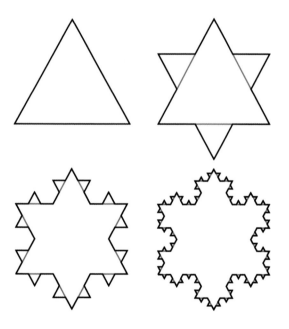

A FRACTAL SNOWFLAKE | This sketch shows the first few steps in the construction of a fractal form known as the Koch snowflake. In this purely mathematical object, each branch holds smaller sidebranches, which hold still smaller sidebranches, all the way down to an infinite number of infinitely small sidebranches.

FRACTALS, CHAOS, AND ORDER

Branched constructions like snowflakes often exhibit *fractal* patterns. The defining characteristic of a fractal snowflake is a *self-similar* structure, where branches have sidebranches, which have their own smaller sidebranches, and so on.

In fact, real snowflakes are only slightly fractal. The first sidebranches rarely have additional sidebranches, so calling the whole assembly self-similar is a bit tenuous. Moreover, noting the fractal qualities of a branched snowflake does not shine any light on its origin. The branching instability is needed to explain why the branches and sidebranches arise in the first place.

On a related topic, the formation of sidebranches sometimes exhibits a chaotic behavior. When the humidity is high and growth is rapid, even the smallest surface imperfections can trigger the branching instability. When the ice growth is especially hurried, sidebranches appear rather chaotically as the crystal develops.

You can see the result of this chaos if you take a close look at a fernlike stellar dendrite, as these crystals typically display erratically spaced sidebranches. Not only are the sidebranches arranged differently on the six main branches, they may not even match up on the two sides of a single branch. Rapid growth enhances branching and reduces faceting, resulting in a more irregular snowflake construction.

Chaos and order are both present during snow-flake growth, and this is what makes snowflake patterns so intriguing. By itself, the branching instability brings chaos—the unbridled creation of structural complexity, as exemplified by the random sidebranching in a fernlike stellar dendrite. Faceting, on the other hand, brings order, as embodied by the simple perfection of the hexagonal prism. Bring these two forces together, however, and beautifully intricate, symmetrical snowflakes result.

HASTILY GROWN | The sidebranches on a fernlike stellar dendrite sprout chaotically and are not symmetrically placed. In this example, even the two halves of a single branch do not match up.

COMPUTATIONAL SNOWFLAKES

s it possible to "grow" a snowflake on a computer? The answer is yes, although snowflakes in the virtual world are still not quite the same as those in the real world.

The branching instability was discovered in 1963 by William Mullins and Robert Sekerka (it is often called the Mullins-Sekerka instability), sparking efforts to develop computer models of dendritic structures such as snowflakes. Crystal growth behaviors are important in metallurgy and semiconductor manufacturing, so there was considerable interest in modeling branching and related growth phenomena in order to understand them better.

Developing computer models that exhibited both branching and faceting turned out to be a difficult problem. Branching without faceting was worked out by the mid-1980s, yielding dendritic structures with rounded edges. Surprisingly, until just a few years ago, no one had ever made a physically realistic computer model that looked like an actual snowflake.

The breakthrough came in 2005 when mathematician Clifford Reiter established that cellular automata models of diffusion-limited growth could produce structures that were both faceted and branched, reproducing many features seen in snowflakes. Subsequent work has demonstrated

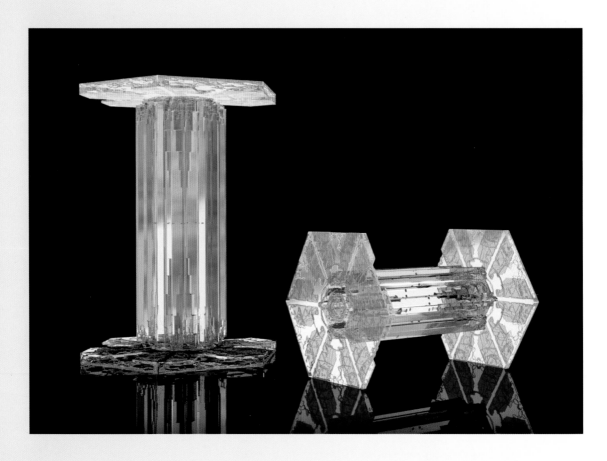

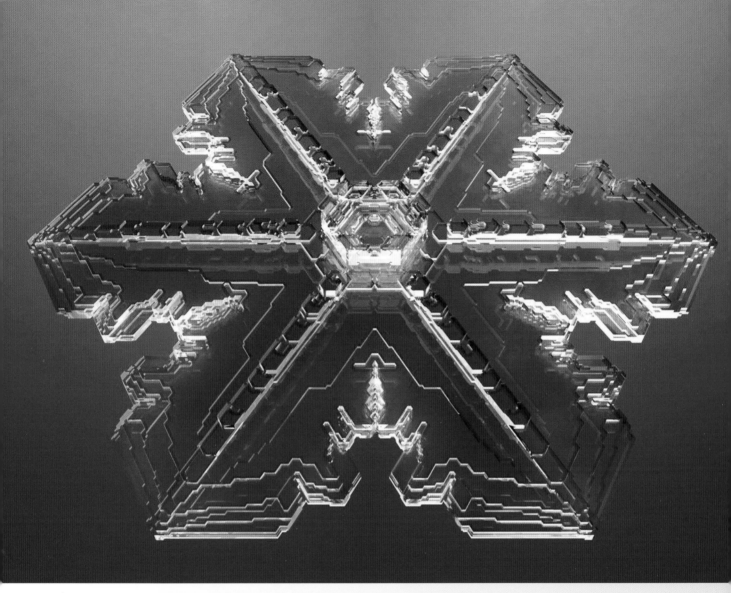

COMPUTATIONAL SNOWFLAKES | These images show a stellar plate (above) and two views of a capped column (opposite) that were "grown" on a computer using algorithms developed by mathematicians Janko Gravner and David Griffeath. The numerical models were turned into color images by Antoine Clappier. In both cases, these virtual crystals resemble natural snow crystals.

stellar dendrites, capped columns, hollow columns, double plates, and other snowflake types.

The computer modeling problem is not solved yet, however, as a number of serious discrepancies remain between real and virtual snowflakes. Adjusting the computer models to produce observed snowflake structures seems to require physically inaccurate inputs, and this issue has not yet been resolved. For now at least, our computer models do not faithfully reproduce natural snowflakes. Here again, this facet of the snowflake story continues to unfold.

A REMAINING MYSTERY

Faceting and branching explain much of what we see in snowflakes. The interplay of these two processes explains everything from faceted hexagonal prisms to fernlike stellar dendrites. The six-fold symmetry of a snow crystal and the synchronous growth of its six branches are both explained as well. The construction of each snowflake reflects a quiet clash between order and chaos that plays out within the winter clouds.

But there remain some unfinished parts of the snowflake story—chapters that Ken and other scientists are still trying to write. Although we have a good handle on faceting and branching, we still do not understand why these processes change with temperature, as illustrated in the snow crystal morphology diagram. We still cannot explain why snow crystals grow into broad stellar dendrites at some temperatures while growing into slender ice columns and needles at other temperatures. Our scientific journey reaches a substantial roadblock when we come to the morphology diagram.

Ukichiro Nakaya's experiments more than seventy-five years ago revealed how snowflake shapes change with temperature, but observing is not the same as understanding. Nakaya could not explain his morphology diagram, and no one else has been able to explain it either. Why do small temperature changes have a large impact on growth morphology? This is one of the questions that keeps Ken going back to the snowflake lab.

As scientists, we find that our inability to explain the snow crystal morphology diagram is a tad embarrassing. If you want to know about black holes, quantum field theory, Big Bang cosmology, the human genome, genetically modified life

forms—no problem. We have all those covered. But why do snow crystals grow into columns or plates depending on temperature? Nope, that one is too tough. Here we are, well into the twenty-first century, and we still cannot explain a phenomenon as seemingly simple as the growth of ice crystals into plates or columns.

The principal obstacle to understanding the morphology diagram can be found at the ice surface. Snow crystal growth is governed by how quickly water molecules attach to ice surfaces, and that involves the collective motions of many particles. Tracking the multitude of molecular interactions at the ice surface quickly becomes something of a computational quagmire. Figuring out the interaction of two water molecules is doable, but even state-of-the-art computers cannot handle many molecules at once with sufficient accuracy.

In 1611 Johannes Kepler set aside his problem of explaining snow crystal symmetry, because he realized that the science of his day could not provide an explanation. It took another three hundred years before crystallographers solved that problem once and for all, giving us the ice crystal lattice with its hexagonal symmetry. Likewise, we must now look to the future to solve the problem of the snow crystal morphology diagram. Our understanding of the molecular dynamics of crystal growth is not yet adequate to provide an explanation. We hope we will not have to wait another three hundred years.

Modern science, fortunately, is moving rapidly on the molecular front. We are confident that the morphology diagram will someday be a solved problem. But until that day arrives, a bit of mystery remains in the formation of a snowflake.

POND CRYSTALS

Some mornings you can find branched ice crystals growing on the surfaces of still ponds or lakes. This photo shows Rachel holding an example she found on the shore of Lake Superior. The temperature had dipped just a bit below freezing the preceding night, so the ice grew—but not too quickly—across the water's surface. This particular ice sheet shows a branched structure similar to that found in stellar dendrite snow crystals.

Although branching in snowflakes arises from water vapor diffusion through the air, branching in these pond crystals results from heat diffusion through the water. The angles between the branches and sidebranches again reveal the hexagonal structure of the ice crystal lattice.

When the weather is too warm for snowflakes, you might find us searching the shores for exceptional pond-ice specimens. Nature is quite generous in providing beautiful ice phenomena.

CHAPTER SEVEN
Designer Snowflakes

"The scientist does not study nature because it is useful; he studies it because he delights in it, and he delights in it because it is beautiful."
——HENRI POINCARÉ, *SCIENCE AND METHOD*, 1908

Much of what we know about snowflakes comes from watching them grow in the laboratory. Although we cannot monitor the formation of an individual snow crystal as it travels through the clouds, we can study one held captive in a jar. Observing synthetic snowflakes under the microscope, we can witness the development of facets and branches, measure growth rates under controlled conditions, and examine how morphologies differ with temperature and humidity. We can see with our own eyes how a speck of diamond dust transforms into a resplendent stellar snowflake.

FREEZER SNOWFALL
In the decades since Ukichiro Nakaya created his first laboratory-grown snowflake in 1936, numerous others have furthered the craft. A variety of methods have been invented to better observe and control snow crystal formation in all sorts of environments. Nakaya's rabbit hair is not the technology of choice anymore.

One of the simplest ways to make synthetic snowflakes is to start with a synthetic cloud. We use a basic chest-type freezer to demonstrate this—the kind that opens on top. For best results, ours is empty and painted black on the

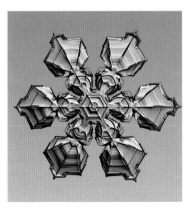  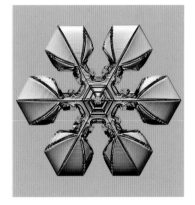

DESIGNER SNOWFLAKES | Laboratory-grown snow crystals such as these can be produced in a great diversity of shapes, like their natural counterparts.

Opposite
ORCHESTRATED SYMMETRY | We directed the shape of this laboratory-grown snow crystal by adjusting the temperature and humidity as it developed.

inside. We have no walk-in cold room in our lab; with our chest freezer and all our other snow chambers, we keep the cold boxed up so we can work in room-temperature comfort.

To start our synthetic snowfall, we open the lid of the freezer and simply breathe down into the cold air, creating a cloud of small water droplets. This cloud is just like the one you make outside when you can see your breath on a cold day. As the cloud hovers inside the freezer, we drop a speck of dry ice into it. This nucleates a few hundred tiny ice crystals, which then grow slowly as they float inside the cloud. Shining a bright flashlight or laser pointer into the freezer reveals a myriad of diamond-dust crystals, sparkling as they tumble.

This kind of freezer snowfall was first demonstrated by Vincent Schaefer in 1946 at the General Electric Research Laboratory in Schenectady, New York, and it yields only small snowflakes. The crystals become heavier as they grow larger, and within a minute or two they fall to the bottom of the freezer. At that point a typical crystal is not much larger than the diameter of a human hair. The sparkle is easily visible with the naked eye, but to see individual crystals well, one needs a microscope.

These minute "snow flecks" may be difficult to see, but they are well suited for our scientific investigations. They are easy to make and are often shaped like plain hexagonal plates and columns. Simpler shapes like these are easier to measure, easier to model on the computer, and their growth is easier to understand all around. In science, it's a good strategy to tackle the simplest problems first, because they are already plenty difficult.

We always seem to be finding new uses for these tiny ice crystals. By examining how they grow at different temperatures and humidity levels, for example, we have studied several aspects of the morphology diagram. And we have been making precise measurements of growth rates to compare with theoretical models of crystal growth dynamics. We use these small crystals to seed other experiments as well, again looking at growth behaviors in detail under well-controlled conditions.

Opposite

GROWN IN A FREEZER | These composite images show tiny snow crystals grown using the freezer method, demonstrating the changes in crystal morphology with temperature. The different panels show plate-like snow crystals grown at 28 degrees Fahrenheit (−2 °C) (top), columnar crystals grown at 23 degrees Fahrenheit (−5 °C) (middle), and larger plate-like crystals grown at 5 degrees Fahrenheit (−15 °C) (bottom).

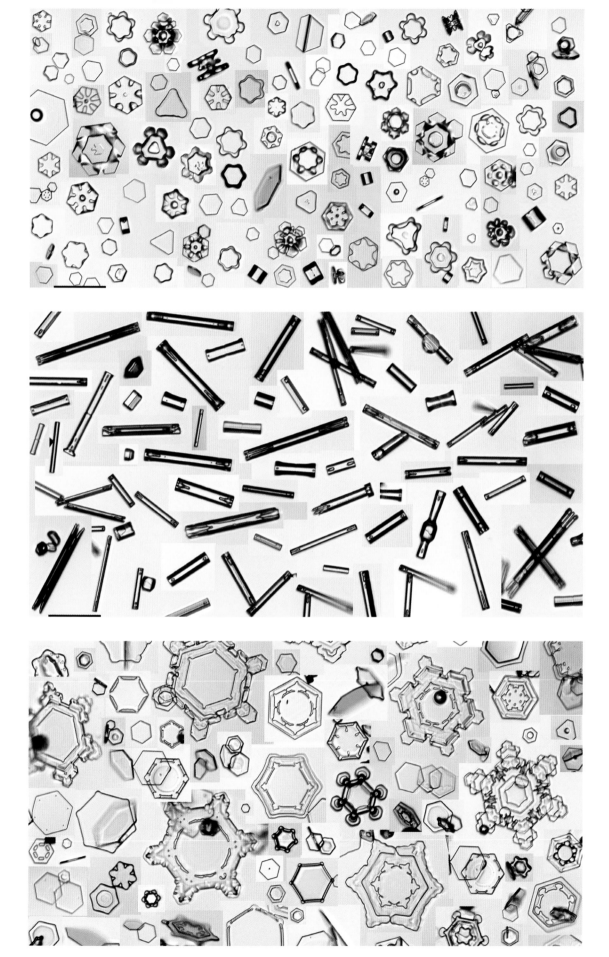

ELECTRIC SNOWFLAKES

Another neat trick for producing laboratory snowflakes is to grow them on the ends of electrically induced ice needles. The basic idea is to use strong electric fields to modify the ice growth behavior, a technique that was discovered in 1963 by Basil Mason and his collaborators at Imperial College London. Ken and his colleagues made some advances in this method as well, discovering how to influence the process using chemical additives and developing a mathematical model of the mechanism underlying electric needle growth. We like to think of these slender needle crystals as the latest high-tech upgrade to Nakaya's rabbit hair.

The method begins with a cluster of tiny frost crystals on the end of a wire in the middle of a chamber filled with cold, moist air. We apply two thousand volts to the wire, and the frost almost seems to spring to life as needle-like ice crystals begin growing rapidly outward. The voltage produces strong electric fields on the crystals, and the fields attract water molecules in the air, greatly accelerating the crystal growth. These ice needles are

extraordinarily slender; near their tips, they are from ten to one hundred times thinner than a human hair.

After the electric ice needles have grown to a desired length, we remove the high voltage so the crystal growth goes back to normal. With a bit of care, we can grow many different snow crystal morphologies on the ends of slender ice needles. Doing so allows us to measure growth behaviors accurately over a broad range of temperatures and humidity levels. The ice-needle technique is especially well suited for experimenting with high-humidity growth forms not encountered in nature.

SNOWFLAKE ON A STICK | The photograph on the upper left (opposite page) shows several electrically induced ice needles growing out from a thicker wire covered with frost. The other photos show examples of different types of snow crystals growing on the ends of electric ice needles.

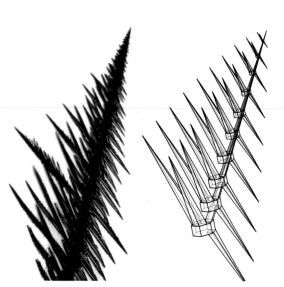

FISHBONE ICE | This photo (left), along with the corresponding sketch (right), shows an unusual type of dendritic structure that we call a fishbone crystal. It grows when the temperature is near 23 degrees Fahrenheit (−5 °C) and the humidity level is exceptionally high. Although a common denizen of our experimental chambers, this ice crystal morphology does not normally occur outside the laboratory.

CRAZY CRYSTALS | This photo shows numerous fishbone crystals growing on the tips of snow stars that formed on the ends of electric needles. We encounter some bizarre ice constructions in the snowflake lab.

Opposite
NOVEL DESIGN | An unusual sequence of growth conditions was used to create this peculiar laboratory snow crystal. It is unlikely that you would find a crystal like this floating down from the clouds.

CURIOSITY-DRIVEN RESEARCH

People often ask us about practical applications of this research, or more generally why we are doing it and what it might be good for. For the most part, our snowflake studies are an example of what is called curiosity-driven research, motivated by a basic desire to understand how things work. Snow crystals simply fall out of the sky with these amazing patterns and structures, and modern science still cannot explain their formation in detail. With more than seven billion people on the planet, surely a few of us can be spared to look into these matters.

We also like to point out that this is not a large or expensive research effort. Ice is easy to make in your kitchen freezer, and growing laboratory snowflakes does not require enormous resources. So if you are imagining a large team of scientists in white lab coats, all bustling about in well-lit rooms full of shiny new equipment, well, that is pretty far from reality. Picture instead Ken sitting in the corner at a messy lab bench, carefully constructing an electronically controlled freezer system, occasionally enlisting help from an interested student or two. Also, you can rest assured that none of your tax dollars have gone into our snowflake projects.

On the science side, Ken likes to quip that he is just a middle-aged guy with an unusual scientific hobby. People have been trying for centuries to figure out how snowflakes work, and the job is not finished yet. We are striving to add our own small contributions to a subject of scientific inquiry that has been ongoing for more than four hundred years.

CLOCKWORK BRANCHING | We induced periodic sidebranching in this remarkable snow crystal by cycling the humidity at regular intervals. We have never seen a natural snow crystal with such precisely ordered sidebranching.

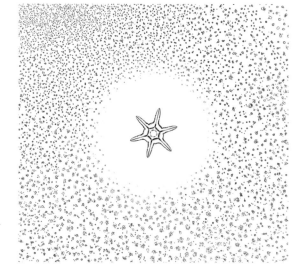

CLOUD ON A GLASS PLATE | Blowing moist air onto a cold glass plate produces a coating of fog droplets, which appear as small specks in this laboratory photograph. The growing snowflake absorbs water vapor from the air, so nearby droplets evaporate away. This snow crystal is just 0.4 mm (0.015 inches) across.

SNOWFLAKE ART

When we want a break from the science, we like to think about snowflakes as a purely artistic pursuit. Our outdoor adventures into snowflake photography are one result, and we have been creating and photographing snowflakes in the lab as well. Although we can capture some gorgeous snowflakes falling from the clouds, making our own means we can document individual crystals as they grow, watching each intricate design as it unfolds.

When crossing over from science to art, we like to call our creations *designer snowflakes*, because we can dictate their final shape by controlling the growth environment. The process becomes a new kind of ice sculpture, in which we use the physical rules of faceting and branching to fashion distinctive crystalline structures. Best

of all, we can make and photograph our own snowflakes all year-round, right here in sunny Southern California.

Beginning with one of our freezer snowfalls, we let some tiny crystals drop onto a glass plate, positioning an isolated hexagonal prism directly under our microscope. Blowing moist air gently down onto the plate causes the stationary crystal to grow larger, and we watch while it does. Water droplets often fog up the glass as well, creating what can be thought of as a two-dimensional cloud, constrained to the surface of the plate. Just as it happens in real clouds, the nearby droplets evaporate away to feed the growing crystal.

Interestingly, these snowflakes do not grow on the surface of the glass plate but slightly above it. The seed crystal develops into a mushroom-like structure—a tiny ice pedestal holds up a large,

SURROUNDED BY FOG | As this stellar snowflake grew larger, the nearby liquid droplets evaporated, maintaining a fog-free zone next to the crystal.

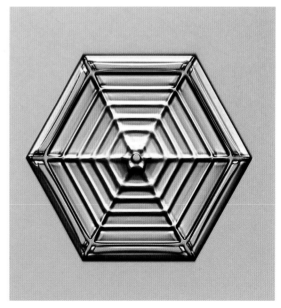

GEOMETRICAL PRECISION | The spider-web pattern seen in this small snowflake was created by applying periodic changes in temperature as the crystal grew. The hexagonal rings of ice are a bit like the rings of a tree, recording the history of the crystal's growth.

CROWDED CRYSTALS | These three snow crystals grew in close proximity as they rested on a glass plate. They interfered with one other's growth as they competed for available water vapor. As a result, the branches between them are shorter than the outer branches.

thin, platelike crystal. Because the snowflake is not actually touching the glass (except at its center), it grows nearly the same as if it were freely floating. As with rabbit hair and electric needles, this mushroom structure is yet another trick for growing suspended snowflakes in the lab.

By changing the temperature of the glass plate and the humidity in the air above it, we control the way each crystal grows and develops. It takes typically thirty to sixty minutes to produce a large stellar snowflake, and we marvel at each emerging detail revealed by the high-power microscope. Adjusting the growth environment allows us to design each snowflake in real time as we watch. We have created our own little snowflake factory.

High yield is not a selling point for this factory; our rate of manufacture is about one snowflake per hour, and that's on a good day. It would take a lifetime to produce enough to make a single snowball. But what we lack in quantity, we make up for in quality.

Up in the clouds, snowflakes have a tough and uncertain life. They are constantly exposed to harsh winds, collisions with cloud droplets and other crystals, and the constant threat of degradation by evaporation. By the time they reach the ground for us to observe and photograph, snowflakes often appear a bit travel-worn, if not worse. The battering they take in the clouds is one reason well-formed, symmetrical crystals can be quite difficult to find in the wild.

Back in the lab, on the other hand, we can create nearly ideal growing conditions, with none of the trials and tribulations of life in the clouds. When the machinery is all working as it should, this allows us to produce remarkably pristine crystals.

We are still perfecting our snowflake-making techniques, learning what we can and cannot do

CRYSTAL CLEAR | We illuminated this designer snow crystal from the side to give it a glassy look.

FROZEN JEWEL | It took over an hour to grow the broad sectored plates on this snow crystal.

with our crystals. It feels a lot like riding along inside a cloud while watching a single snowflake form, except we can control its growth by simply turning some knobs. We can reproduce many features seen in natural snowflakes, plus we are able to create unusual growth conditions not found in nature, such as high humidity levels or electric fields, or unusual background gases and chemical additives. We cannot guide the ice into just any shape, as we are bound by the rules of crystal growth. But that still leaves many possibilities to explore.

The artistic side of our snowflake making has been a nice complement to the science, plus the art has been paying the bills. We are fortunate in that we have been able to support our snowflake activities primarily through sales of photographs and books (thanks for buying this one!), so we keep marching on. It is an odd hobby, to be sure, but one we find most satisfying.

IDENTICAL-TWIN SNOWFLAKES

Occasionally we will find two seed crystals that dropped side by side onto our glass plate, giving us the opportunity to watch their simultaneous growth. If the seeds land too close together, then each interferes with the other, yielding two misshapen snow crystals. But if they are sufficiently separated, each grows essentially independently of the other.

As we change the temperature and humidity around them, each adjustment alters the way the

IDENTICAL-TWIN SNOWFLAKES | This photograph shows two snowflakes that grew side-by-side on a glass plate in our laboratory, both surrounded by a field of water droplets. Since they grew under nearly identical conditions, the two crystals are nearly indistinguishable.

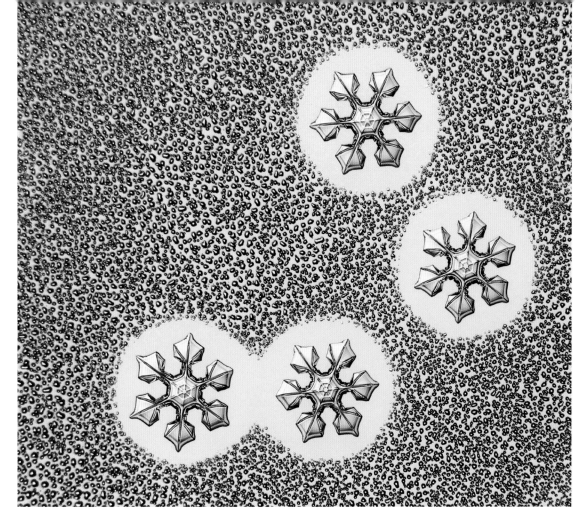

SNOW CRYSTAL QUARTET | This photograph shows four snowflakes on a glass plate, surrounded by a field of water droplets. Since they grew simultaneously, all four crystals are similar in appearance. However, you can see that the two lower crystals were beginning to interfere with one another, slightly stunting the growth of the branches between them.

branches grow, and that means both snowflakes grow in synchrony. The end result is a pair of quite similar crystals that we call *identical-twin snowflakes*. As with identical-twin people, our crystal twins are obviously very closely related, just based on their appearance, but they are not exactly alike in every minute detail.

So how does this affect that old saying that no two snowflakes are alike? That adage is still valid when applied to natural snowflakes, but it does not work so well for synthetic snowflakes. Up in the clouds, the growing crystals are in constant motion through a turbulent atmosphere.

During the half-hour or so a crystal is forming, it experiences hundreds or thousands of chaotic twists and turns in the wind. And each minor course change affects the final shape of the crystal, because the growth behavior is so sensitive to temperature and humidity. The probability that two natural snowflakes would experience precisely the same changing conditions at precisely the same times is vanishingly small.

Our laboratory twins are closely matched because the two crystals remained stationary, both fixed to the glass plate, as we changed the temperature and humidity around them.

No laws of nature were violated here; instead we found a loophole. The laboratory simply provided a more controlled and predictable environment than that experienced by natural snowflakes up in the clouds.

Of course, the question of identical snowflakes is always one of degrees. If you look carefully at our doppelgangers, you will see that they are not exactly identical. With additional time and effort, we could surely improve upon this. The more precisely we can engineer a uniform growth environment, the more precisely alike the final crystals will be. Absolute perfection is impossible but we can get pretty close. If there were a compelling reason to do so, we could gear up our snowflake factory to churn out—with robotic precision—a whole series of nearly identical snowflakes. How dreary that sounds. Let us be thankful that Mother Nature adds such a delightful element of randomness to her snowflake making.

LABORATORY CREATIONS | Small changes in temperature and humidity yield large changes in how snow crystals grow. We exploited this fact to create these designer snowflakes with a variety of patterns.

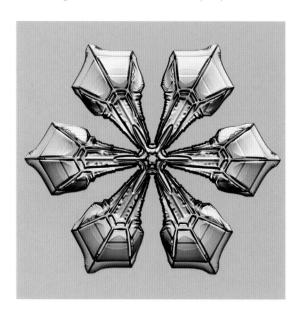

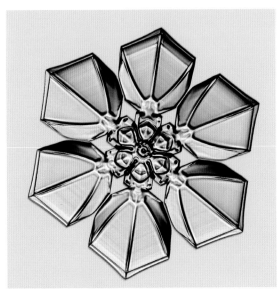

CHAPTER EIGHT

Snowflake Photography

"Besides combining her greatest skill and artistry in the production of snowflakes, Nature generously fashions the most beautiful specimens on a very thin plane so that they are specially adapted for photomicrographical study."
—WILSON BENTLEY, *POPULAR MECHANICS MAGAZINE*, 1922

Snowflake photography has a lot in common with other forms of nature photography—it requires an artistic eye, some suitable optical gear, and a compelling desire to just go out there and take some pictures. The craft presents its own challenges as well, in that snow crystals are small, a bit fragile, prone to evaporation and melting, and—as if that weren't enough—they need to be handled outside in the wind and cold. We have been photographing the little guys for many years, and we have developed a few techniques along the way. In this chapter we describe how we took the pictures in this book, and we have a look at a sampling of gorgeous images from the growing community of snowflake photographers.

THE GREATEST SNOW ON EARTH

Living in Southern California, we frequently embark on what we call our *snowflake safaris*—traveling to distant locations in pursuit of especially photogenic snowflakes. Although this may be an unusual use of one's vacation time, people often journey from the frozen north to take in the sights and beaches of Southern California, so maybe a few of us should make the trek in the opposite direction. Also,

photographers frequently travel great distances to find scenic places and exotic wildlife, and we feel that snowflakes should receive the same consideration. Assuming, then, that one is inclined to embark on a snowflake safari, the first question becomes choosing a location. Some places are clearly better than others, but where are the best spots? Where does one find the greatest snow on earth?

If large, well-formed stellar snowflakes are the goal, then the morphology diagram points to temperatures near 5 degrees Fahrenheit (–15 °C). But stellar crystals are only part of the snowflake story, so what about capped columns or other interesting crystal types? Are mountain snowflakes the same as those falling at low elevations? Do different types of snowflakes appear in different regions of the globe?

In fact, no one really knows the answers to these questions. Although we have much meteorological data on snowfall amounts around the globe, the mere appearance of falling snow does not guarantee exceptional snow crystals. Quantity of snow and quality of snowflakes are, if anything, inversely correlated—light snowfalls tend to have better snowflakes, while heavy snowfalls are often wet and gloppy, and blizzards are just a windy mess.

Opposite
COLORFUL SNOWFLAKE | We found this snow crystal in Fairbanks, Alaska. It was photographed using colored lights to accentuate the internal patterns in the ice.

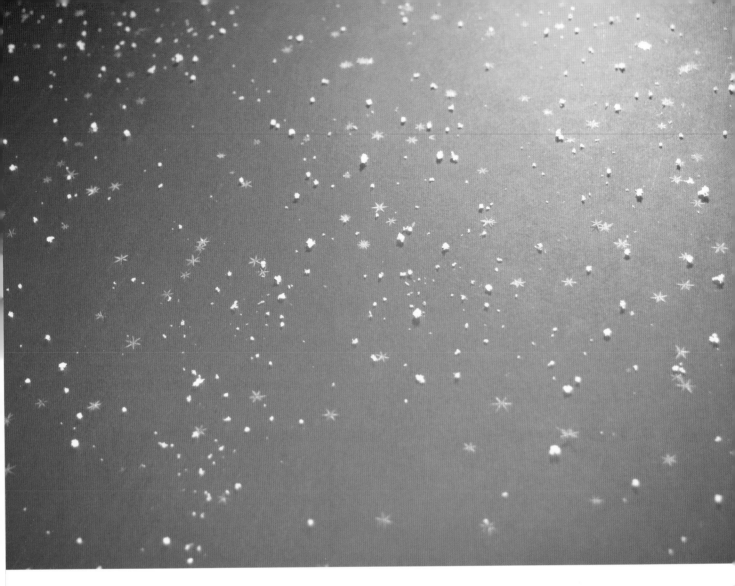

IN THE FIELD | The picture on the right shows Ken outside photographing snowflakes, while the above image shows a batch of crystals on the collection board.

BIG-GAME HUNTING

When we see larger snowflakes appearing with some regularity, we usually stop catching crystals directly on glass slides. We switch instead to a much larger collection board so we can look over more snowflakes quickly. In keeping with our snowflake-safari theme, we switch to a strategy we call *big-game hunting*.

We prefer blue foam-core cardboard for our collection board, as ice crystals stand out clearly against its smooth, dark surface. As the snow is accumulating on the board, we scan back and forth, up and down, looking for promising specimens. From time to time we brush the snow off as the surface becomes cluttered. The big-game crystals are large enough that we can select them mostly by naked eye or perhaps using a low-power magnifier. Being nearsighted is somewhat advantageous during these searches.

A practiced eye is remarkably adept at spotting exceptional snow crystals on a crowded collection board. It's a bit like searching for edge pieces when you first start a jigsaw puzzle; your brain just focuses on finding those straight edges while ignoring all the curvy pieces. For snow crystals, we focus on faceted surfaces that show a bit of sparkle. Often we work in the dark (the nights are long in winter), and then a spotlight is helpful by providing bright reflections from the facets.

With a good-sized collection board, we can look though many snowflakes quickly and easily. At any given time, there might be upward of a thousand crystals on the collection board, and in less than a minute we can usually pick out the best subjects. Doing the math, this means we have actually sifted through millions of crystals over the years in our quest for photographic subjects. Some of our snowflakes really are one in a million.

When we spot a promising snowflake, we pick it up off the collection board and transfer it to a glass slide for photographing. We do this using a small paintbrush, rolling the bristles gently under the crystal to lift it without damage. People often think that handling a tiny, fragile snowflake must require the skill of a surgeon, but it's not terribly difficult. The crystals are stronger than they look, and most survive the transfer step undamaged.

A bigger annoyance is the wind. When a snowflake is perched on our paintbrush, the slightest gust of wind will . . . whoosh—off it goes. That irreproducible snowflake pattern, whatever it was, is lost forever. Fortunately, irreproducible does not mean irreplaceable. We simply go back to the collection board and select another.

As soon as a snowflake is on the glass slide, it goes under the microscope for a closer look. If what we see under magnification leaves us cold (as it were), we might skip the picture. Perhaps the crystal is rimed, misshapen, or just not so interesting to look at. More often, however, we are stirred to adjust the lighting, adjust the focus, and take the shot. After recording its image in the camera, the lovely snowflake is unceremoniously tossed onto the ground with all its brethren.

Melting and evaporation are additional problems we have to contend with when photographing snowflakes. Heat from our hands or the lights will cause a crystal to start evaporating. The outermost regions evaporate away most quickly, and the crystal slowly disappears before our eyes. On warmer days, the crystals may simply melt into tiny puddles on the glass. The solution to these problems is just to work quickly. We try to pick up each crystal and photograph it in less than a minute.

Finally, there is the cold. Handling the paintbrush, the glass slides, and the camera

continued on page 132

EVAPORATION VERSUS MELTING

The first column of photographs on the opposite page shows a snow crystal slowly evaporating away under the lights of the microscope. From top to bottom, a period of two minutes elapsed. You can see how the finer features and extremities are the first to disappear, leaving behind a simpler shape.

The other column of photographs shows a crystal melting. The temperature was just below freezing, and just twenty-seven seconds elapsed from top to bottom. It can be quite a challenge to photograph snowflakes in such warm conditions.

A snow crystal is always changing, and it begins evaporating as soon as it stops growing. If the clouds are high in the sky, the crystal may experience significant evaporation before it even reaches the ground. The crystal shown below arrived in a partially evaporated state, looking quite travel-worn.

On the ground, snowflakes continue to change, and they tend to lose their finer features quickly. After a day or two in a snowbank, the crystals are mostly blocky in form, with little resemblance to their initial elaborate structures.

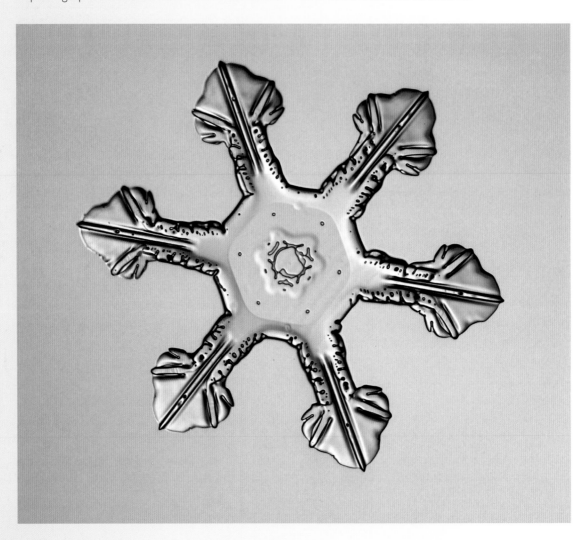

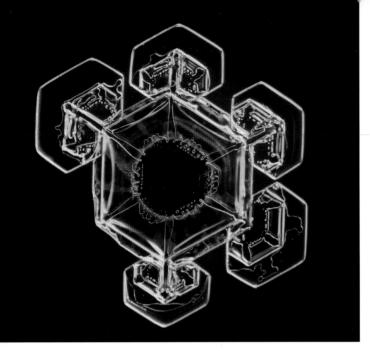

BROKEN SNOWFLAKES | We do sometimes damage snow crystals during handling. Fortunately, there are plenty more.

continued from page 129

requires some dexterity, so we have found it necessary to work with bare fingers. Standing around sifting through snowflakes one by one can be a challenge when the temperature is around 5 degrees Fahrenheit (–15 °C).

THE SNOWMASTER 9000

The question of photography gear is always one of degrees. Good equipment is expensive, but owning the very best is not necessary for taking fine pictures. The quality of the equipment is not as important as the eye of the photographer—and a little luck helps too.

For our photographs, we wanted high magnification to capture even the smallest snow crystals, and we were after bitingly sharp images that showed every intricate detail. After trying a few different approaches, we ended up building a special-purpose photomicroscope optimized for snowflake photography—a piece of hardware we dubbed the *SnowMaster 9000* (not to be confused with the Snowmaster snow scoop, the Sno-Master ice shaver, the SnoMaster fridge freezer, Snow Master winter tires, or even the Snowmaster 9000 tool in World of Warcraft. We came up with the *SnowMaster* moniker in pre-Internet times, before realizing how many others had done the same. But the name stuck, so here we are).

Basically our photomicroscope is a set of three microscope objectives covering a range of magnifications attached to a digital camera using a long extension tube. The objectives are mounted on a custom-built turret, making it easy to change magnifications for different snowflakes. Our specimens are almost always on glass slides, and these rest on a translation stage that moves up and down to focus, as is typical with microscopes.

During one of our snowflake photography sessions, a snowmobiler sharing our motel came over to chat and have a peek through our microscope. He got a kick out of seeing snowflakes this way and asked how much this contraption of ours must have cost. When we told him (some thousands of dollars), he could scarcely believe that anyone would put down that kind of cash to photograph snowflakes. We had noticed, meanwhile, that he was driving a fully equipped Hummer pulling four shiny new snow machines. He was a bit taken aback when we pointed out that his entertainment bill was actually quite a lot higher than ours. It all depends on where you put your priorities; to each his own.

BENTLEY'S DILEMMA

How one illuminates a snowflake to photograph it requires more care than you might initially think. The difficulty originates from the fact that snowflakes are made of clear ice, and photographing transparent objects can be a challenge.

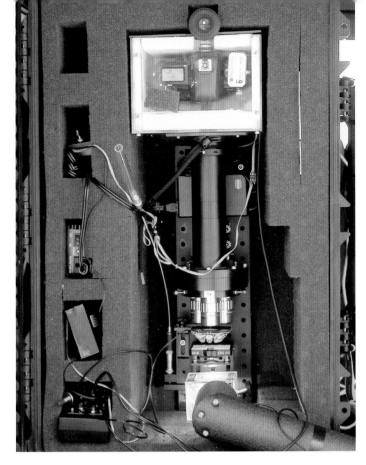

THE SNOWMASTER 9000 | Our set-up for photographing snowflakes is basically a camera, a set of microscope objectives, and a lamp. The assembly is mounted in a suitcase for portability.

We refer to this as *Bentley's dilemma*, because Wilson Bentley faced the illumination issue as soon as he took his first snowflake photograph.

One straightforward approach to illumination is simply to shine light on a snowflake from the side or above, as you would anything else. This is how you see a snowflake on your sleeve, after all. At high magnification, however, crystal edges scatter a lot of light and appear bright, while thin, plate-like regions scatter very little and show up dark. The high contrast between different parts of the same crystal is dramatic but tends to obscure fine detail and may look harsh.

Bentley avoided the sidelighting issues by illuminating his crystals from behind, so his microscope was looking at light transmitted through the clear ice. This kind of backlighting introduced a different dilemma, however, in that now the contrast in the image is quite low. Imagine photographing a sliver of glass placed on a white piece of paper, and you can see the problem: the clear snow crystal appears white against a white background.

Bentley solved his dilemma by modifying a copy of the image after the picture was taken, removing the bright background and replacing it with black. This was long before Photoshop, of course, and even before photographic film, so he accomplished the task working directly on glass photographic plates. For each of his thousands of snowflake pictures, Bentley made duplicate negatives and then painstakingly scraped the photographic emulsion from each plate outside the crystal perimeter with a sharp knife, a process that took about an hour per picture. A print made from such a "blocked" negative yielded a bright snow crystal against a black background. Bentley's white-on-black approach accentuates the outline of a crystal but makes it difficult to see internal structural details and surface patterns.

SIDE ILLUMINATION | This photograph shows a snow crystal illuminated from the side, as it would appear on a black surface. You can see that the smooth parts of the crystal are clear, while the edges scatter light and thus appear white.

BACK LIGHTING | This photograph shows the same snow crystal illuminated using colored lights from behind. This type of lighting accentuates the three-dimensional character of the crystal.

RHEINBERG ILLUMINATION

Recognizing the pros and cons of side and backlighting, Ukichiro Nakaya found that oblique illumination—from behind but off to one side—provided an effective alternative. With oblique illumination, the ice crystal acts like a piece of shaped glass that bends the incident light. This adds a sense of depth to the photographs, giving an overall better view of the internal makeup and surface patterning in snowflakes.

We take oblique illumination a step further in our snowflake photography by applying what is called *Rheinberg illumination*. First described by London microscopist Julius Rheinberg in 1896 for biological specimens, this technique shines different colored lights in from different angles, essentially giving a combination of several colors of oblique illumination simultaneously. Different hues then highlight different parts of a crystal, providing an additional dimension of color to snowflake photography.

THE RIGHT MOMENT

Nature photography is often about being in the right place at the right time. If you want to photograph a brilliant rainbow, a spectacular lightning display, or any number of other natural phenomena, the trick is being there and ready when it happens. In contrast, if you want to take a group photograph of your family, you are free to pick any convenient time. Assembling everyone together and getting them all looking decent for just one measly picture may be maddeningly difficult, but at least it is theoretically possible.

When seeking outstanding snowflake photographs, we have to be ready and waiting when outstanding crystals are falling. This means we are at the mercy of the weather, over which we have no control and only a little ability to predict. Picking a convenient day is meaningless, because there might not be any falling snow that day, perhaps for hundreds of miles in any direction. If the clouds are not doing their part, then one simply cannot be a snowflake photographer that day.

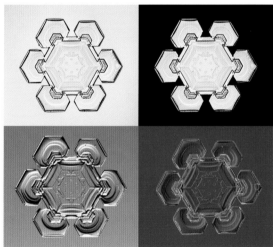

ILLUMINATING SNOWFLAKES | These four photographs show the same snow crystal using different illumination techniques. The first image used plain back lighting, resulting in a rather flat view of a white crystal against a white background. Digitally altering the photo to make the background black gave the second image, which is similar in appearance to Bentley's photographs. Rheinberg illumination adds a greater sense of depth in the third image, while using a rainbow of colored lights yielded the fourth photograph.

RAINBOW LIGHTING | The flamboyant photograph, top, was taken using Rheinberg illumination with a rainbow of colored lights. People often think we used polarizing filters here, but this is not the case. Nor were the colors painted on digitally. This is pretty much how the picture came out of the camera. The smaller photograph above shows the same snow crystal, except using more muted colors.

INEXPENSIVE MACRO | This photo shows the year stamped on the surface of a US penny; it was taken using a smartphone with an $8 clip-on macro lens. One does not necessarily need a lot of fancy equipment to photograph small objects like snowflakes.

During one of our trips to northern Ontario, a Toronto TV reporter found out we were photographing snowflakes and decided it would make a good story for an upcoming evening broadcast (slow news day, no doubt). The reporter told us that he required authenticity in his reporting, so staging a photo shoot would not work; we had to be out there photographing real snowflakes. In his next breath he then asked whether he could bring his film crew next Tuesday at three o'clock. The poor fellow was disappointed when we pointed out that it might not be snowing that day or any other day we scheduled a week in advance. The weather was not that predictable. After some additional discussion, he decided that snowflake photography might not be such a great story after all.

POINT-AND-SHOOT SNOWFLAKES

It is not necessary to own a lot of expensive gear to take terrific pictures of snowflakes. High resolution optics are expensive, but high resolution is not the only consideration in snowflake photography. Sacrificing some resolution gives a greater depth of field, for example. Much can be done using a handheld, point-and-shoot camera with some macro capability.

With a handheld camera, no snowflake handling is necessary. One can simply set out a collecting surface and photograph snow as it lands. Moreover, the collecting surface can be as varied as one pleases, for example a scarf, mitten, stone, leaf, or piece of wood. There are many possibilities for intriguing surfaces.

Freshly fallen snow resting on the leaves of the trees provides an opportunity for composing a classic winter scene. Nature does the hard work of growing the snowflakes and setting them out on display. The real challenge is dragging yourself out of bed on cold mornings to find these idyllic photo opportunities.

IN THE EYE OF THE BEHOLDER

After nearly two decades of photographing snowflakes ourselves, we enjoy all the different styles and techniques being developed by a growing number of snowflake photographers. Some methods are better suited for producing bitingly sharp images that show every nuance in the crystal design, while other methods yield wonderful views of snowflakes in natural settings. Sidelighting reproduces the view you see with a snowflake on your sleeve, while oblique illumination from behind accentuates structural details.

With the advent of digital photography, snowflake photographers in the twenty-first century have been pushing the craft in new directions. Modern cameras have much higher sensitivity than those of even the recent past, allowing more latitude for capturing scenes with low light levels. Photo-sharing websites such as Flickr have brought together people interested in snowflake photography to share new ideas and photographs.

Snowflake photography can be a most satisfying pursuit. Although standing outside in the cold photographing small slivers of ice may not be everyone's cup of tea, perhaps you will relish the craft as much as we have. With new aesthetic directions and ever-improving camera technology, there is a world of photographic expression that remains relatively unexplored. We expect to see much more as people continue capturing and recording these enchanting crystals.

SPARKLING SNOW CRYSTALS | Canadian photographer Don Komarechka captured these two images in Barrie, Ontario, using a handheld camera with a high-resolution macro lens and a ring flash. The front illumination yielded bright crystals set against a dark background. When Don is shooting snowflakes, he takes about one hundred to two hundred pictures of each subject in rapid succession, moving his camera back and forth to obtain good focus over different parts of the crystal. Later he uses image-processing software to combine thirty to fifty of the best pictures into a single, high-resolution image. This high-tech extension of point-and-shoot photography produces exceptionally sharp images that reveal even minute details in the ice. *Don Komarechka*

SNOWFLAKES IN THE FAMILY | Russian photographers Alexey Kljatov and Olga Sytina are keeping snowflakes in the family—Olga and Alexey are mother and son. Both used handheld cameras with macro lenses to capture these images. The pictures were taken in Moscow under ambient light conditions, so a steady hand was required to focus on the small snowflakes. Alexey notes that excellent crystals like these appear only infrequently. Snowfalls bringing large numbers of well-formed crystals might happen only a few times each winter. *Alexey Kljatov and Olga Sytina*

Elizabeth Akers

Josh Shackleford

Jessica Dyer

Marc Kohlbauer

Delena-Jane Lane

ALL ABOUT SNOWFLAKES | The photographs on these two pages show snowflakes in natural light, as they might appear on your sleeve or on other surfaces. They suggest the endless possibilities for capturing snowflake images.

Pam Eveleigh

Jill Lian

Jackie Novak

Out of the bosom of the Air,
Out of the cloud-folds of her garments shaken,
Over the woodlands brown and bare,
Over the harvest-fields forsaken,
Silent, and soft, and slow
Descends the snow.

—HENRY WADSWORTH LONGFELLOW, *THE COURTSHIP
OF MILES STANDISH, AND OTHER POEMS*, 1858

Epilogue

This brings us to the end of our story of the art and science of snowflakes. Along the way, we have examined different types of snowflakes in detail, scrutinized their symmetrical structures, and investigated how such elaborate designs are created within the clouds. We have paused briefly to take an up-close look at the remarkable architecture of these tiny crystals of ice.

From this day forward, consider yourself an authority on all matters pertaining to this most fascinating form of frozen precipitation. For now you too can recognize the curious capped column or majestic stellar dendrite when one lands on your sleeve, and you too can explain the origin of its distinctive shape and symmetry.

In our own lives, we watched our children grow up as we shared with them our obsession with the many facets of snowflakes. Ken's research into snow crystal science has become something of a family enterprise, sending us on snowflake safaris to three continents. Our quests for unusual and beautiful snowflakes spanned those golden years of family adventure, when the kids were old enough to travel but young enough to still want to accompany us. After nearly two decades of travel and study, we still delight at the sight of an extraordinary snow star.

Our journey began with what seemed like a simple question: how do snowflakes form? That question led to Ukichiro Nakaya, the morphology diagram, and our family's first encounter with a capped column. Our laboratory experiments have brought us to creating designer snowflakes, giving us a fascinating new perspective on nature's frozen artistry. As we approach the 130th anniversary of Wilson Bentley's first snowflake photograph, we are now exploring ways to create novel types of snow crystals in a controlled environment, delivering them to the watching eye of our camera as they form and develop in real time. Along the way, we are refining our method for growing identical-twin snowflakes.

We are especially excited about recent developments in the growth of virtual snowflakes on the computer, as these are showing increasing fidelity to natural crystals. This aspect of the snowflake story is changing rapidly, as new computational tools are being developed. Soon comparisons between laboratory snowflakes and their virtual counterparts will add much to understanding the science.

Emerging from the lab, we are planning additional travels with the SnowMaster 9000, revisiting our favorite snowflake spots and exploring new ones. Antarctica and Siberia are high on our list.

We hope that our telling of the snowflake story has inspired you to go outside and take a close look for yourself at the falling crystals. We hope you can find time to pause, take in the calm of a quiet snowfall, and watch your misted breath float upward to join the clouds. And, most of all, we hope you continue to find joy and wonder in the snowflake.

Opposite
INTRIGUING SNOWFLAKES | The more we learn about snowflakes, the more we want to know. This "negative" image shows an exceptional laboratory-grown snowflake.

INDEX

Page numbers in italics indicate an item that appears in a photograph or caption.

SURROUNDED BY SNOW | Rachel and Ken, out enjoying the snowy woods.

ABOUT THE AUTHORS

Kenneth Libbrecht is a professor of physics at Caltech in Pasadena, California, where he studies the molecular dynamics of crystal growth, especially how ice crystals grow from water vapor, which is essentially the physics of snowflakes. He has authored several books on this topic, including *The Snowflake: Winter's Secret Beauty*, *The Art of the Snowflake*, and *Ken Libbrecht's Field Guide to Snowflakes*.

Rachel Wing is a park ranger for the city of Monrovia, California, at the foot of the San Gabriel Mountains. With her background in geology, she specializes in balancing wilderness preservation with wildfire safety for nearby residents. Rachel loves to hike and climb in these hills, especially on the rare occasions when snow graces the desert-like plants. She has been accompanying Ken as a snowflake chaser for nearly twenty years. Rachel and Ken live in Pasadena, and they have two children.